BEEN THERE, DONE THAT

BEEN THERE, DONE THAT

A Rousing History of Sex

Rachel Feltman

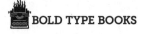
BOLD TYPE BOOKS

New York

Bold Type Books
116 East 16th Street, 8th Floor New York, NY 10003
www.boldtypebooks.org
@BoldTypeBooks

Printed in the United States of America

First Edition: May 2022

Published by Bold Type Books, an imprint of Perseus Books, LLC, a
subsidiary of Hachette Book Group, Inc. Bold Type Books is a co-publishing
venture of the Type Media Center and Perseus Books.

The Hachette Speakers Bureau provides a wide range of authors for speaking
events. To find out more, go to www.hachettespeakersbureau.com or call
(866) 376-6591.

The publisher is not responsible for websites (or their content) that are not
owned by the publisher.

Print book interior design by Amy Quinn.

Library of Congress Cataloging-in-Publication Data
Names: Feltman, Rachel, author.
Title: Been there, done that: a rousing history of sex / Rachel Feltman.
Description: First edition. | New York, NY: Bold Type Books, 2022. |
    Includes bibliographical references.
Identifiers: LCCN 2021057368 | ISBN 9781645037163 (hardcover) | ISBN
    9781645037170 (ebook)
Subjects: LCSH: Sex—History. | Sex customs—History.
Classification: LCC HQ12 .F445 2022 | DDC 392.6—dc23/eng/20220203
LC record available at https://lccn.loc.gov/2021057368

ISBNs: 9781645037163 (hardcover), 9781645037170 (e-book)

LSC-C

Printing 1, 2022

CONTENTS

Contents

EVERYTHING WEIRD IS NORMAL—EVERYTHING NORMAL IS WEIRD

THIS IS A BOOK ABOUT SEX. ALL SORTS OF SEX. SEX BETWEEN PEOPLE with mutually ambiguous genders. Sex between gangs of female chimps. Sex between slutty amorphous blobs in the primordial ooze. Sex between you and your hand. All kinds.

But I just keep spiraling back to duck penises.

During the process of lovingly crafting the tome you're about to read, I learned that only 3 percent of male birds have dicks. This information should not have come as a shock to me. For starters, how often do you see a bird with a boner? Not often, I hope. Also, I should know as much: I'm the executive editor of a science magazine, and I collect random facts about sex like it's my job (it sadly is not). And yet.

Here's the thing: penises are not a given. Birds, reptiles, amphibians, egg-laying mammalian weirdos, and some fish eschew the genitals we humans think of as traditional in favor of the "cloaca," a sort of all-purpose, on-off ramp into and out of the body.

Animals with cloacas use them for pretty much every sort of in-take and outtake, save for eating and breathing—though some turtles do take gasps of air through these Swiss Army buttholes, when the mood strikes. Chickens also lay eggs through their asses, and that's all there is to it. Sorry for ruining your breakfast, but the world is built on shit and blood, and you're going to have to accept that eventually. It may not be pretty, but the cloaca is a magnificently efficient orifice.

The existence of this multihyphenate hole is not news to me. Still, I was *also* vaguely aware that male cloacas contained some-thing phallic hidden within: a furled member that sat poised to emerge for intercourse.[1] I suppose I've always assumed it must serve as a passable penis once deployed. Reader, it does not.

In point of fact, a male bird's equipment—when gently coaxed from its cloacal comfort zone by the hands of a well-meaning biologist—looks something like a tiny larva. Often no bigger than a sliver of newly cut fingernail and shining with the half-baked, gooey look of a fragile embryo, these wee appendages serve only to sputter out seminal fluid. The delivery mechanism is a phallus-free "cloacal kiss," wherein birds just . . . touch their respective parts together.[2] This sometimes lasts for mere fractions of a second, with nary a thrust. In any case, penetration is so far off the table it might as well be on the floor.

It had somehow never occurred to me that male birds don't all have respectably sized penises (proportionate to their puny bird bodies, of course) tucked away in their all-purpose poo-and-pee chute.

I blame ducks.

Ducks—along with geese, swans, and a number of flightless birds such as ostriches and emus—make up the proud phallic 3 percent of the avian world. But they don't just have penises. They have horrifying penises. They took the concept of a dick and ran with it headlong into the gaping maw of hell. Male ducks possess massive, spring-loaded, corkscrew-shaped members (sometimes barbed, always ugly) that can, in some cases, stretch as long as their bodies.[3] A duck penis is a sight to behold and, once beheld, impossible to erase from your memory.

There's an interesting evolutionary story behind these horrid lengths. See, birds that pair up monogamously without fail tend to have short phalluses, if they have them at all (most, of course, do not, but my point here is to make it clear to you that swans have really tiny dicks). Ducks, in contrast, do the worst kind of playing the field. They often engage in "forced copulation"—rape, in human terms—and that's probably why they've evolved to have such relatively massive dongs.[4] And also why those dongs are curled up tight like springs. And sometimes barbed.

See, the female duck has evolved tricks of her own to help her dictate which male will father her offspring.[5] Her vagina is basically a haunted funhouse. It's like the Winchester Mystery Mansion in there. Twists. Turns. Cul-de-sacs. Stairways to nowhere. I'm not kidding.

When a female duck is chill and cooperative (dare we say, consenting enthusiastically?), it's simple enough for a male to get his penis into the actual vaginal canal so he can successfully inseminate her. But if she struggles, there's a chance that he'll wind up slipping through the biological equivalent of a trap door and spilling his seed into a dead-end pouch of skin. Forced copulations are still common, and no doubt lead to plenty of pregnancies, but this physiological quirk gives females a shot at, well, calling the shots.

A 2007 study led by Patricia L. R. Brennan put forth the idea that ducks have evolved in a sort of sexual arms race, with vaginas growing increasingly baroque while male penises get bigger and twistier to try to compensate, and this view is now widely accepted by people who think about duck penises for a living.[6]

It's easy to ignore or forget the fact that duck genitals aren't bizarre just because of their size and generally threatening vibes. Considering that 97 percent of birds do perfectly fine sans penis, it's strange that duck penises exist at *all*. Scientists now believe that the common ancestor of all cloacal animals possessed a phallus, but clearly most descendants found the extra appendage increasingly useless. Some ancestral duck with a real toxic attitude shot bravely in the opposite evolutionary direction, presumably mocking its increasingly dickless cousins in the locker room while its own reproductive organ morphed into an eldritch horror.

Compared to the theatrical shock of duck sex, the sticky business of cloacal kissing seems like a bit of a letdown. Giant, horrible penises are just bound to get more airtime. Somewhere along the way, I guess I just internalized the assumption that birds must have penises—however puny those penises might be. I unconsciously assumed that penises defined sex, even in the animal world.

Our assumptions aren't all that different when it comes to human proclivities.

Most of us are taught, at varying ages and to varying degrees, that "sex" happens when a penis makes its way into a vagina. We might have some pushback from more progressive people in our peripheries, but that doesn't change the forcefulness of the predominant messaging. All sorts of people can have sex in all sorts of ways, we "know"; but advertisements, media, textbooks, and religious institutions repeatedly tell us that the default setting is a man with a penis lying atop a woman with a vagina, often under sheets that seem to magically keep his butt covered.

Those of us who fancy a bit of variety in our definition—whether in terms of the sexy acts being committed or the sexy people committing them—are constantly pushing back against this long-held sense of the status quo. But the status is *not* quo. The world of sex, at large, has very little to do with what we humans consider normal. In fact, the animal kingdom and all of human history are full of scenarios just like the antipenile revolution in the sky.

I don't mean to suggest that we should remodel our society to reflect the smooth and utterly dickless copulatory practices of most birds. But take a closer look at evolutionary history, at other animals, at our own past, and at different cultures around the world, and you'll find that sex, on the whole, is a lot weirder than you might think.

And that might kind of be the point of sex. It's an elastic, adaptable, catch-as-catch-can, DIY-friendly, totally open-source method of reproduction and social connection that keeps the world as we know it spinning. It cares not for the fate of your penis. The dinosaurs that didn't go extinct became birds, we all know, and birds with dicks, in most cases, became birds without.

If humanity still exists in a few million years, who's to say we won't go the way of our feathered friends?

Sorry, I don't mean to threaten the evolutionary future of your genitals. My point is that what is normal right now has not always been normal; nor will it always be so. From the biological realities of our carnal acts to our generally acceptable methods of courtship, the world of sex is the furthest thing from static. Instead, sex is a shimmering spectrum of colorful moving targets.

Before we dive in, here's a quick disclaimer: I'm not going to literally attempt a speed run of all of history to get you from point A (the dawn of dicks) to point Z (the four hours it took the TikTok algorithm to diagnose me as a bisexual cis-woman who's kinda "meh" about gender and married to a nice man with floppy hair). There's just too much.

I'm also not a historian, let alone one specializing in the history of sex or in queer studies; I'm more of a great-at-finding-fun-facts-for-cocktail-parties girl, but in a professional capacity.

There are plenty of books out there in which people smarter and more interesting than I am have outlined portions of sexual history for you, in detail, and you can check those out to immerse yourself in, say, how same-sex attractions affected troops in World War II, or what queer cinema can tell us about France in the 1960s, probably, or how Dolly Parton became a gay icon. You can read books by evolutionary biologists to give you a crystal clear picture of our best guesses and latest theories about how nature invented dicks (duck and otherwise). You can read books by sexual health experts about the complete care and keeping of vaginas throughout recorded history. There are entire books dedicated to the practice of grafting monkey testicles onto male humans for the purposes of giving them more reliable boners.

This is not one of those books. This is a book by someone who decided to cram the entirety of the history of sex into roughly three hundred pages. But—you may ask—why?

I'm writing this because I wish I'd read it a decade ago. I've put a lot of work into understanding sex, personally and professionally—probably more work than most people have the time or resources to put in. I still feel like I've got a veritable dusty attic's worth of hang-ups to unpack. My relationship with sex and my body is a work in progress. I'm writing this because I hope you can learn quickly what it took me far too long to learn: that today's mainstream definition of sex is deeply flawed and that this has the ability to cause us harm. Sex can be anything and everything humans—and birds, and Neanderthals, and slime molds—dream or want it to be.

Here's what you need to know. *Whatever* sex you can imagine today? Humans have been there before. Animals have done that and are still doing it. The whole world, I promise you, has been there and done that. And you can too!

I know one thing for certain: getting some context for how topsy-turvy our modern-day sense of what sex should and shouldn't be was a crucial first step in my journey to being a happy and healthy human. So I come to you humbly offering a smattering, a taste, a mere assortment of amuse-bouches of sexual expression and queer existence and horny exuberance through history.

Sex has always been weird. Sex has always been normal. All that's changed is how we talk about it.

From duck penises and koala chlamydia to gay cowboys and Lysol douching, we'll peel back the layers of everything your high school biology and history textbooks got wrong—and, more importantly, everything they failed to even mention. Hold on to your cloacas.

Chapter One

WHAT THE HECK IS SEX?

· ·

Where amorous bonefish from the ancient world
give us a glimpse at the early days of boot-knockin'.

· ·

HERE ARE FIVE LESSONS I HOPE TO HAVE TAUGHT YOU BY THE END OF
this book:

1. Our earliest ancestors may have been queer as all heck,
   and a lot of cowboys were gay.
2. People have *literally always* wanted to bone as much as
   possible without having babies and were willing to shove
   crocodile dung up their hoo-has to do so.
3. That weird thing you like? It's fine. I promise. Like, really. It's probably not even that weird. Like, not to offend
   you? I'm sure you're a unique snowflake and all, a real
   rebel without a cause, but, like, *trust me*, people have been
   weirder.

4. People have been conspiring to make you think you masturbate too much since the 1800s, if not earlier.
5. There are sexually transmitted infections that are *good* for you, and most of the ones that are *bad* for you still aren't nearly as bad as you've likely been led to believe.

We're going to get into all of this and a whole hell of a lot more to boot. But before we can talk about chastity belts, the surprising number of secret sex museums in European history, giraffes peeing on each other in a horny way, and mail-order radium suppositories, we need to answer one teeny-tiny question: What *is* sex, anyway? And to answer that query, we need to go back a couple billion years.

There was a time before sex. When the earth was new, all living things reproduced asexually: rather than finding sexual partners, individuals begot copies of themselves to perpetuate their ilk.[1] This was simple. It was efficient. Every member of the species was capable of reproducing and did so without help from any of their kin. Life boiled down to eating, avoiding being eaten, and occasionally copy-pasting your DNA by splitting yourself in two. Some prokaryotes learned to swap DNA with one another on the fly, which helped their species adapt and combine genetics in new ways. But their offspring were still the result of whatever genes progenitors had handy at the time—not of a dalliance with another individual.

Sometime around one to two billion years ago, as best as the fossil record can tell us, the first eukaryotic organism decided to muck about and make things a lot messier.[2] This common ancestor, likely a single-celled protist, maintained the ability to clone its own cells—in a way, we're all reproducing asexually every time we make new cells within us, which translates to nearly four billion births a second per person—but it also started making sex cells, or gametes.[3] (To complicate matters, some researchers argue that this last eukaryotic common ancestor, or LECA for short, was actually not a single cell containing all the genetic traits necessary to make a eukaryote, but rather a population of diverse single-celled organisms that swapped *just* the right genes at *just* the right time to make all the proteins it takes to build a defined cellular nucleus.[4] Luckily for us and for the length of this book, we don't need to know which scenario is correct in order to know that eukaryotes . . . happened.) Unlike the so-called diploid cells that each contain the entirety of an organism's genetic code, gametes are haploid, which means they only carry half. They need to combine with other haploid cells to create a fully functional set of chromosomes.

Bangiomorpha pubescens, so named as the first known occurrence of "sexual maturity" for life on earth, is currently considered to be the oldest fossilized organism that certainly had these abilities.[5] The specimens in question are thought to be just over a billion years old, making them the most ancient fossilized critters that appear to be single, complex organisms—as opposed to colonies of unicellular bacteria.

Unlike those of earlier organisms, *Bangiomorpha pubescens* spores show three distinct morphologies, representing cells it could have used to reproduce asexually, but also "male" and "female" cells similar to those used for sexual reproduction in modern *Bangio* algae. It seems likely now that all extant eukaryotes, or organisms with cells divided into membrane-bound organelles, have sex in their ancestral history—even the few that reproduce exclusively asexually today. They may have come from lineages that dabbled in both modes of procreation, then reverted (sort of like how whales and dolphins descend from animals that emerged from the sea, evolved into mammals, and then scooted their way back into the ocean for reasons unknown).

But this wasn't quite *sex* as we know it. Plants reproduce sexually by trading pollen on the breeze. Our first sexually reproducing ancestors likely just oozed up against one another at the cellular level. When did we start, you know, *doing it*?

Our oldest evidence of penetrative intercourse is about 385 million years old and comes in the form of fossilized remains of the way too aptly named *Microbrachius dicki*. I know. I know! But believe it or not, *M. dicki* got its rather pointed moniker from Scottish baker-turned-botanist Robert Dick in the nineteenth century.[6] Mr. Dick would never know that the ancient armored fish he chiseled free from rock were sexual revolutionaries. Not until 2014 did a study confirm that the remains showed the earliest known example of internal fertilization and copulation. And oh, did that reveal some glamorous origins for getting it on: *M. dicki*'s eight-centimeter-long body included a "bony L-shaped genital limb" called a clasper, which males used to transfer sperm to females. Not to be outdone, the species' better half developed "small paired bones to lock the male organs in place."

But the first known instance of sex as we know it still wasn't really *sex as we know it*. Based on the placement of those bony, interlocking claspers, paleontologists say the frisky fish probably swam side by side to do the deed. "With their arms interlocked," one of the lead study authors said in 2014, "these fish looked more like they are square dancing the do-se-do rather than mating."[7]

I can't speak for everyone, but I'd say we've come a long way. And our journey from square dancing in the sea to engaging in the thrilling array of activities we now call sex is full of shocking, disturbing, and hilarious twists and turns. *M. dicki* is just the world's introduction to intercourse; once humans hit the scene, we really learned to have fun with it.

But enough about *how* sex came to be. Why did we start having it? The answer may not be as straightforward as you think.

A CONFUSING STRATEGY AT BEST

"We do not even in the least know the final cause of sexuality; why new beings should be produced by the union of the two sexual elements," Charles Darwin wrote in 1862. "The whole subject is as yet hidden in darkness."

We've learned a lot since then, but we're still stumbling around in want of a proverbial flashlight. The question of why we have sex might seem to have an easy answer, but researchers are debating this great conundrum to this day.

An academic paper titled "Classification of Hypotheses on the Advantage of Amphimixis" may not sound like a thrilling read, but I promise it's worth a bit of your time.[8] "Amphimixis" is a fancy term for sexual reproduction (so fancy, in fact, that I learned it myself about five minutes ago), and the point of this 1993 study is that theories on the evolutionary purpose of sex were so numerous as to require organization. Even some thirty years ago, all the prominent inklings we had floating around about why sex might exist numbered in the dozens. Sex remains a mystery to this day.

One easy explanation for why sex took the world by storm—and the reason you probably learned in school, if your teacher broached the subject at all—is that sex creates genetic diversity. By creating kiddos with half of each parent's DNA, which allows for countless potential genetic variations with each coupling, a sexually reproducing species can quickly proliferate an army of unique snowflakes. Genetic diversity is, of course, beneficial. If something goes wrong—a famine or a pathogen or any number of other inevitable disasters—having more types of individuals in play on the board ups your chances of *someone* having what it takes to survive. If your entire species is cloned from one common ancestor and that common ancestor doesn't happen to have built-in resilience against the new plague in town, your entire lineage is toast.[9]

That's what happened to bananas: Central American growers once relied on the Gros Michel cultivar's ability to reproduce asexually to help them fill grocery stores with fruity clones. Then, in the late 1800s, a fungal plant pathogen called *Fusarium oxysporum* f.sp. *cubense* hit the scene.[10] Gros Michel bananas were totally susceptible to fusarium wilt, and because the entire world supply, at least when it came to large-scale farming, was just a copy of the same banana, there was no hope that a few hardy individuals with useful genetic quirks would survive the onslaught. By 1950, most bananas being sold in stores were instead clones of the Cavendish banana, which was the closest growers were able to get to the taste of the Gros Michel while still producing a fusarium-resistant crop. Predictably, this is going to backfire on us any day now—the pathogen has mutated to target the Cavendish. Whoops!

But sexual reproduction isn't as much of an obvious win as you've likely been led to believe.

Consider how much energy goes into sex. It requires that two individuals find each other and go through some kind of information exchange. It means that not all of your genes will make it to the next generation. Indeed, each sexual encounter risks your *best* genes being arbitrarily left in the dust while chromosomes that leave you vulnerable to disease and other pitfalls propagate. It doesn't matter how molecularly resilient your ancestor was to one environmental hazard or another if you end up with the wrong combo of DNA on the relevant stretch of your genome.

And it's not as if an asexual species is *truly* stuck with only the genetic material of its ur-ancestor. Mutations can occur by chance all the time, some of which will aid survival and thus be passed on due to the mutated party's relatively long life, creating subtle differences between various clonal lineages.

Evolutionary biologists' best guess is that sexual reproduction is simply a more efficient way to handle both good and bad mutations.[11] That is, sex makes it easier to flush out genetic oopsies without losing entire lineages of the species (for example, someone with a not-awesome genetic predisposition for dealing with some particular environmental scenario may have some offspring that get more advantageous genes from the other parent instead). Moreover, in an asexual lineage, beneficial mutations show up one by one and can only be transferred if organisms happen to meet up and exchange information (like when bacteria learn how to avoid antibiotics by swapping DNA with others they come into contact with). By contrast, sexual reproduction makes it easy to create offspring with all the best new bits and bobs of DNA found in the collective grab bag of the species. This also means that there's less likelihood of various distinct beneficial mutations creating competition between members of what once used to be a single, cloned species.

We see evidence for either or both of these hypotheses in so-called facultative organisms. Some creatures—mostly plants, but also some animals—can switch between asexual and sexual reproduction as needed. Many of them engage in what's called condition-dependent sex, where their method of reproduction shifts based on certain environmental conditions. In general, it seems like such critters are apt to flick the sex switch into the "on" position when the world around them is changing in dangerous ways. The wee crustacean *Daphnia magna*, for instance, is more likely to do the nasty when food is scarce or temperatures are in extreme flux.[12] When conditions are peachy keen, these plankton produce only "females" and reproduce by cloning. Similarly, the rodent parasite *Strongyloides ratti* switches to sexual tactics when confronted with a strong immune response from its host.[13]

In general, it seems that sexual reproduction is a shortcut to the kind of diversification that can help a species thrive. The "Red Queen" hypothesis—so named for Lewis Carroll's *Through the Looking-Glass*, where Alice realizes she must run as fast as she can merely "to keep in the same place"—suggests that life on earth is so competitive that organisms need to keep their feet on the gas pedal of evolution just to have a shot. In other words, defaulting to asexual reproduction might mean your species misses out on a genetic opportunity that could otherwise keep it alive in the face of an unknowable future threat.

But the examples highlighted above also make one wonder why we—and so many other eukaryotes, which make up all living things save for bacteria and archaea—got stuck with sex as a full-time occupation. Why wouldn't we have maintained the option to bud in a pinch, if our genes are hardy and mates are hard to come by? That seems to be the case for several species of shark, females of which are able to execute asexual "virgin births"— more accurately known as parthenogenesis—when males are scarce and conditions are cozy.[14] If fish can take matters into their own hands, why can't humans do the same? Even if we're closer to understanding why sex happens in the first place, the reason why sex, once evolved, tends to become a species' exclusive reproductive strategy remains murky.

One of the most perplexing aspects of doing the deed is that it often (though not always) means that a species will be split into different sexes, only some of which have the ability to create new offspring. In theory, a single hermaphroditic phenotype could handily cover both sides of the sexual equation. So why waste energy on having males at all?

WHY EVEN ARE MEN?

The question of why males even exist is an ongoing and completely straight-faced field of study. Still, I'll grant you that it sounds like more of a misandrist Twitter rant.

Indeed, when I wrote about a study examining this conundrum in 2015, I published it under the headline "Scientists Examine Why Men Even Exist," and I knew *exactly* what I was doing.[15] The Tweets I got in response were not very nice, but the study was: biologists at the University of East Anglia spent six to seven years observing two sets of around fifty generations of beetles to try to suss out whether sexual selection might be the key to it all.

Let's rewind briefly to Darwin. When he wasn't busy having kids with his first cousin,* Darwin posited that something called sexual selection might be key to reproductive success. His more famous theory of natural selection held that only the most genetically fit individuals would survive long enough to reproduce, thereby making their DNA more likely to persist through the generations. But his theory of sexual selection granted that some qualities might dominate the gene pool not by nature of their inherent superiority but thanks to the ability of individuals to be picky about their sexual partners. Sex doesn't just give you more opportunities to create combinations of genes that benefit your species; it also means you have to be "good" enough—by some definition or another—to get to reproduce. No sneaky budding shortcuts for you!

Since male beetles can't make their own babies when times get tough and have little to do with raising offspring, the researchers in that 2015 study wanted to see whether the benefits of sexual selection might be enough to justify their existence. They

---

* All kidding aside, research on the subject of reproduction between first cousins suggests that, while it's not ideal from a genetic-diversity standpoint, it's not the worst thing in the world. But this becomes less true the more often your family line dips back into itself, as evidenced by many royal families in less genetically literate eras, since your pool of shared versus unique genes gets smaller and smaller with each criss-crossed branch of the family tree. So it's a seal probably better left unbroken, if your heart can help it.

removed selection from the equation for some couples by randomly pairing them up into monogamous sets, then upped the selection quotient for other females by placing them in groupings with increasingly extreme sex ratios—capping off with habitats featuring just ten gals with ninety potential mates to pick from.

Almost a decade later, they tested the genetic resilience of those two groups of beetles by inbreeding test subjects with their siblings. No matter which experimental lineage they came from, forcing brother and sister beetles to mate amplified any genetic mutations they'd acquired over the generations (this is why you don't want to have kids with your close relatives, folks, and the kind of strategy that gave the Habsburgs their signature jaw, though Darwin clearly didn't get the memo). In the groups where sexual selection had been weak or impossible, offspring started dying quick by the tenth cycle of inbreeding. But beetles descended from those able to practice sexual selection survived as many as twenty generations of sibling incest. Sexual selection, it seemed, gave them more to work with when times got tough.

This reminds us all of the importance of not looking for potential co-parents at family reunions. But the study also offered a small but compelling piece of evidence for the importance of choice in giving sexual reproduction its power. And so, choice could explain why an entire sex evolved seemingly for the sake of being chosen as mates.

From the 2015 study, it seems that perhaps males can serve as a sort of evolutionary doodle pad, where the species can give risky genes and behaviors a chance to stay around and prove their worth. A certain number of females must have DNA that allows them to safely come of age, have sex, procreate, and care for their offspring. That's hard work, over a long-term horizon, and really stifles creativity—you don't want to risk loads of potentially good, but potentially bad, mutations in that pool. So having another sex creates the opportunity for those nongestating individuals to live fast and die young, biologically speaking. A male beetle, for example, only has to live long enough to get lucky once to share some of his more beneficial genes, making it less of a tragedy if some of his other mutations are more detrimental.

But while we currently have more questions than answers about the existence of men, one thing is clear. The paradigm of "male" and "female" organisms butting uglies is an arbitrary way for reproduction to have panned out. Sex could have gone differently in any number of ways. And in many cases, it still does.

Chapter Two

HOW NORMAL IS HETERONORMATIVITY?

· ·

In which we learn how grandmothers and gay uncles are essential to the survival of the species, visit the ancient escapades of totally-not-gay men around the world, ask how American bison got to be so bisexual, and ponder if queerness is where it all began.

· ·

PICTURE THE AMERICAN WILD WEST—THE GREAT FRONTIER. MANIFEST destiny and colonialism as far as the eye can see. Cows. Corn? I don't know, I'm not a cowboyologist. A tumbleweed goes by. Someone yodels. A man spits tobacco juice on your shoe. Stuff like that.

A group of rugged, rough-and-tumble cowboys eye a herd of bison. It's the 1880s, after all, and bison (that is, the common plains bison of the species *Bison bison*) have been hunted to near extinction.[1] Far from their modern-day status as the national mammal of the United States, the behemoth bison, to a cowboy, is nothing more than a cash cow. Their days are numbered. Do these majestic beasts know what fate awaits them?

Someone spits more tobacco juice on your shoe (probably). The air is tense with anticipation; someone takes aim.

Then, a pause. The gun is lowered. Awkward laughter. Knowing glances between the cowboys. Because one of the bison bulls is apparently feeling amorous. Because it's not a female he decides to try mounting. Because the bison, my friends, is as American as apple pie *and* as gay as a maypole. And these cowboys, believe it or not, may not have been too straight themselves.

Reviewing numerous existing studies on homosexual behavior in American bison, researchers from Belgium in 2006 noted that same-sex mounting had been "regularly observed in bison (*Bison bison*) males" throughout the preceding decades.[2] Moreover, some data even suggested to the researchers that certain age-classes of males indulged solely in same-sex mounting. Thoughts on why this behavior was so common varied across the studies: some concluded same-sex mounting was a form of play or a way of establishing dominance, while others suggested it merely came down to an individual's otherwise scant opportunities to have sex. But there was no debate about whether such couplings took place—all the data suggested they were quite common.

And it's not just bison on the open range. Indeed, male domestic cows are so prone to mounting one another that there's a name for the mountee: the "buller." According to that same 2006 paper, bullers—a term reserved for bulls subject to repeated mountings from multiple peers in short order—tend to be unfamiliar to the group they're in and/or naturally low in the pecking order. They tend to have higher estrogen and testoster-one levels, and it's thought that the addition of growth hormones in feed (along with overcrowding) has contributed to a historical uptick in this phenomenon in the United States. Being a buller isn't exactly a party: repetitive mountings can leave the receiver with an irritated rump, exhaustion and stress, hair loss, and even broken bones. But in the wild world of *Bison bison*, at least, male mountings seem to be a much more benign aspect of life.

When the 2006 researchers collected their own data—from a massive Belgian farm where more than one hundred imported American bison grazed freely year-round—they noted that male homosexual mounting was indeed common. But they found lit-tle evidence of its value in establishing dominance. Instead, they landed on the notion that it probably served either as a form of play or a way of practicing for adult reproduction.

Bison lesbianism, however, proved to be more mysterious.

For starters, it happened more frequently than expected. The authors noted that most of the big studies on bison homosexual-ity treated female-female mounting as something that occurred only in very specific situations. At least one study left ladies out entirely. But it turned out to be about as common as male homo-sexuality. "Although not all females were observed to perform homosexual behavior," the authors noted, "these interactions seem part of the normal female bison ethogram." But when the authors tried to evaluate the purpose of the behavior—by mea-suring how indulging in it seemed to affect a bison's social status, fertility, and so on—they saw only minimal effects. Though the researchers couldn't explain away the incidents as being useful for one reason or another, we know female bison mount female bison just as much as males mount males.

And the men chuckling at those bulls and bullers? Well, some of those boys were known to seek out same-sex action too. Orville Peck and Lil Nas X may have raised the concept of the gay cowboy to an art form, but they by no means invented it.

Historians writing on homosexual behavior among the rugged boys of the Wild West—especially ones commenting on the subject more than a decade or so ago—have been careful to hedge any statements that contradict our heterosexual American ideal. "It's important to know the history of homosexuality," history professor Peter Boag of Washington State University told *True West Magazine* for an article titled "Homos on the Range" back in 2005.[3] "Society didn't really designate people as homosexual or heterosexual through most of the 19th century; it was not really until the 20th century that those identities crystallized." That means that in all-male societies, for example, men commonly engaged in same-sex acts that weren't seen as "gay" in the modern sense. And yet, according to that same article, though they may not have been "gay," these cowboys shared codes to establish preferences, like alluding to an enjoyment of Walt Whitman's work to hint that they shared his orientation. That sounds like a little more than circumstantial palling around, don't you think?

Even in 2005, the author of the aforementioned article was quick to point out that a photo showing, for instance, a room full of male cowboys slow dancing with one another was not necessarily gay, because there were no women around and men just . . . had to . . . slow dance. Still, it is safe to assume that yes, obviously, as it has always been and always will be, some of the cowboys who had gay sex were really, actually gay. And in the strange circumstances of the frontier, it seems, they were allowed to live as they pleased. In fact, the wide-open spaces and mind-your-own-beeswax attitude of the American West, notes Boag, actually seems to have drawn a crowd of gender-nonconforming folks as well, including a few who lived their lives quite out loud, at least relative to the East Coast norms of the day.

And if the American bison and the American cowboy turn out to be as queer as quiche, then what else in our history, our culture, our science, and our world has always been different than we've been told? We know that sex between males and females exists. But so, dear reader, does every other kind. And it has been that way since the very beginning.

⟡

One of my favorite silly little internet jokes goes something like this: *Sorry, you want me to pick a place to eat dinner? I'm a bisexual switch.* I've never made a decision in my entire life.[4]

---

* The *OED* shows that the word "bisexual" was first used to mean an attraction to both men and women in 1906, in an English translation of Otto Weininger's *Sex and Character*. Before that, it was used to refer to organisms or objects with a combination of male and female characteristics. In 2020, *Merriam-Webster* updated its definition of the word to break free from the gender binary; in its official capacity, "bisexual" now refers to people who are attracted to more than one gender or, depending on whom you ask, people who are attracted both to members of their own gender and to people of other genders or no gender. A "switch" is someone who, depending on their mood and the circumstances, may enjoy playing either a dominant or a submissive sexual role. I am the physical embodiment of an UNO wild card, and the world is a smorgasbord of sexual possibility. This mostly just makes me anxious.

This joke is funny because it's true, but also because it's patently false: queer people—myself included—grow up today having to make all sorts of decisions about their sexual presentation. While being some flavor of queer is as natural as breathing (we'll get to the science that says so in a minute), the lines of our world are drawn to be colored in with heterosexual lifestyles. To be gay—by which I mean not-straight, just to ensure we're all on the same taxonomic page moving forward—often means creating rogue masterpieces atop the blueprints laid out for us at birth.

Many people figure, either subconsciously or out loud, that anyone who isn't heterosexual and cisgender operates outside the norm. Even if you don't begrudge the existence and happiness of LGBTQIA+ people—and perhaps even if you've been known to don a piece of gay apparel or two yourself—you might assume that homosexuality is a bit of a paradox. After all, our underlying biological imperative is to pass along our genes and keep them circulating through existence. And a human who only seeks out partners of the same sex is not going to end up making a baby (at least not without medical intervention, which wasn't a thing for most of human history, or a surrogate or sperm donor). So why did we evolve into a species where sometimes people are simply and completely flaming?

As a college professor of mine once bluntly put it in an attempt to fluster us spawnless youngsters, humans are biologically pointless until they have kids. And, you know, there's something to stop and unpack there, because—ew, right? Not cool.

We *know* that there's more to being human than propagating our genes. We live and love and laugh and make TikTok videos and write poetry; we feel and make others feel; we take care of one another; we fuck up monumentally and make war and peace and create culinary abominations like the KFC Double Down. That's what separates us from other animals, or at least most of them. And we've got plenty of people on the planet, so it's perfectly reasonable for some of us to ditch that biological imperative entirely. We're not bacteria in a Petri dish or bunnies in a warren—we've evolved to a point where spreading our seed doesn't have to be life's endgame. Still, we recognize that we wouldn't exist today if an evolutionary hustle to have procreative sex hadn't driven our ancestors.

But a new and intriguing school of thought around the evolution of sexuality suggests a tidy explanation for the apparent paradox of preferences that *seem* to thwart our biology. In fact, according to this nascent line of thinking, being queer hasn't been screwed out of our gene pool because being queer is actually the *default* setting. At least, that may have been the case in some distant and ancient ancestor.

A subtle shift in one's perspective might explain the weirdness of animal sexuality—in both bison *and* cowboys. When complete homosexuality is framed in the *standard* paradigm, Yale University researchers specializing in ecology and evolution noted in a 2019 paper, it comes at a high cost. (That means that being gay makes you less likely to reproduce and spread your genes, not just that people are jerks about it.)

And yet, queerness *does* persist. So maybe the evolutionary cost isn't as steep as it seems. Perhaps being gay isn't actually evolutionarily confusing—perhaps humans are just confused about evolution. The Yale researchers propose a simple change: Stop asking *why* animals engage in same-sex sexual behavior. Instead, start asking why they *don't*.

"In any trait so widely seen across different animal species, you would usually at least consider the hypothesis that the trait was there from the origin," Julia Monk, lead author of the paper and a PhD candidate in forestry and environmental studies at the Yale School of the Environment, told *Popular Science* in 2019.[5]

That means that we too often assume that our first sexually reproducing ancestors must *not* have engaged in same-sex intercourse. Yet we don't actually know that this is true. In fact, it's just as reasonable—if not more so—to assume that our most ancient ancestors actually did the deed quite indiscriminately.

Imagine a primitive and immobile multicellular ancestor of animal life. This progenitor reproduces sexually but has yet to develop anything as elaborate as clear sexual dimorphism (where one sex is easy to distinguish from another), let alone anything as elaborate as a courtship ritual. Unlike a pair of birds displaying plumage and dancing around one another to curry favor, any member of this species may just have attempted to mate with any comrade it came across in the primordial ooze. In this scenario, any sort of pickiness would have put an individual's genetic line in peril.

"The notion that SSB [same-sex sexual behavior] is a recently evolved and distinct phenomenon from 'heterosexual' sex," state the researchers, "is symptomatic of the kinds of binary essentialism that hinder not only social liberation and equity, but also scientific discovery."

Bear in mind that this notion—that the road to today's animal life was paved with organisms that swung both ways—is just a hypothesis. Moreover, we currently have no means of confirming this version of our evolutionary history.

But let's continue to play out the thought experiment. Instead of wondering how homosexuality could possibly persist in the gene pool, we now get to wonder something else: When did having any kind of sexual orientation became the norm? We can picture some early critters starting to develop clear sexual differences to make it more likely for mating to lead to reproduction; when spawning something is your ultimate aim, it's useful to have more than a fifty-fifty shot of the individual you hook up with being reproductively compatible with you. Once this happened, perhaps attraction to the physical traits that indicated an individual belonged to the opposite sex would make your genes more likely to proliferate than if you used the earlier scattershot approach.

But here's the key: if the population is starting from a place of pure pansexuality,[*] this new heterosexual adaptation isn't going to kick in universally, like a flipped switch. Some individuals will still manage to successfully pass their genes on without having a particular attraction to members of the opposite sex or aversion to members of their own. As long as enough babies are being made from generation to generation, there *isn't* actually a steep cost to some members of the species desiring nonprocreative sex.[6]

Let me pause to get one thing straight: I'm not, and that doesn't require biological justification. Even if there is no biological purpose to being something other than heterosexual—even if

---

[*] *Merriam-Webster* defines "pansexual" as follows: "of, relating to, or characterized by sexual or romantic attraction that is not limited to people of a particular gender identity or sexual orientation." Many people use it to signify an orientation unbound from the gender binary. There's a lot of debate and personal preference involved in identifying as either pan or bi; you only need to stress about this if you feel like it.

queerness is truly *brand spanking new*—it's still nobody's business whom you're attracted to. We don't need an evolutionary rhyme or reason for existing as queer people; queer people exist, which means they deserve to exist. But it's worth prying at the cracks of the simple narrative we've been sold about our species' sexual heritage. The question of whether our ancient ancestors might have had sex in ways that surprise us shouldn't be a new one. Unfortunately, the history of science as we know it is rife with researchers imposing their own lifestyles and social norms onto the natural world. As you'll learn in this book, we take for granted countless paradigms that we should have shaken up a long time ago. The assumption that heterosexuality is so necessary and ancient an evolutionary tactic as to make homosexuality puzzling is just one of them.

But nonprocreative sex isn't just a nonliability! In fact, science has come up with a few ways in which being a bit fruity might help a species thrive.

SO'S YOUR GAY UNCLE

You may already be familiar with the so-called grandmother hypothesis. But it's exciting enough to repeat. And before we turn to gayness, it seems, first we need to understand Grandma.

Some researchers aim to explain the continuation of life after menopause in human females with the grandmother hypothesis.[7] See, in most animals that reproduce sexually, your life is pretty much over once you've aged out of making babies. But humans with uteri have a vexing tendency to stay alive long after that whole process has shut down, which usually happens somewhere between the mid-forties and the mid-fifties.

The grandmother hypothesis suggests that in humans—and, weirdly enough, in killer whales, where females also live long past the age of fertility—there may be some benefit to having a period of life not devoted to reproduction. Life after menopause may help protect your genetic lineage by allowing you the time and resources to assist your children with the care and keeping of *their* children. Thus, being a grandmother raises the likelihood of your genes making it into future generations of humanity (or orca-nity, as the case may be).

Human and orca grandmothers saving their species is cool enough. But related to that hypothesis is another, which likewise tries to explain why something *seemingly* biologically unnecessary can be, actually, quite necessary. Just as, perhaps, humanity would be nothing without its grandmothers, so too might it be suffering without its queer people.

The kin selection hypothesis, raised by evolutionary psychologists in 2010 and sometimes referred to colloquially as the gay uncle hypothesis, posits that the nieces and nephews of people who don't have their own kids might benefit from the extra attention.[8] The idea came about as the result of a study on groups of *fa'afafine* in Samoa, who are assigned male at birth but identify as a third gender and typically have relationships with men. In these exact circumstances, the evolutionary benefit of having a queer person as kin is patently obvious; *fa'afafine* live in Samoan communities where families are tight-knit and geographically close, and they tend to contribute time, money, and attention to the rearing of their siblings' kids.

It's unlikely that tykes with queer aunts, uncles, or auncles see much of this sort of benefit in other parts of the world, as industrialization and sprawl have made it increasingly uncommon for extended families to live close enough, physically or emotionally, for a parental sibling's generosity to amount to more than ample chunks of birthday cash. But it was different in ancient times: having babies was both fairly unavoidable in a heterosexual relationship and fairly crucial to keeping one's village afloat, and folks also had to hunt and forage for enough food to fill hungry young mouths. Back then, researchers argue, the world at large looked much more like the communities that feature *fa'afafine* than the towns and cities in which many of us now live—and having an aunt or uncle who helped feed you instead of producing hungry tots of their own may have given you a real edge. The kin selection hypothesis, if correct, would suggest that having some queer people around made family lines more likely to thrive, which would keep that family's genes circulating more widely.

Regardless of whether having a nonreproducing sibling makes you more likely to successfully raise more children, the idea that this explains *why* some of us are queer presupposes that there is some genetic component to sexuality, which is something of an open question. But while being not-straight surely does not come down to one or even a small handful of genes, it does seem likely that DNA is part of the complex equation (along with environment, cultural influence, hormones, and who knows what else, because humans are messy). And if there are, indeed, genetic markers that make one more or less likely to be a friend of Dorothy, they may have been more likely to stick around in the gene pool thanks to helpful gay uncles.

(Note: Even if this effect is real and has had a real impact on our species, it is still possible that your own personal gay uncle is an unhelpful jerk. Moreover, it is okay if you do not choose to use your own gay uncle powers to help your nieces and nephews and other niblings become prolific breeders.)

Another idea that I find particularly fun looks to the benefits of same-sex attraction in other mammals; in many primate species, for instance, being open to taking a tumble with a member of the same sex can really help grease the wheels of social interaction.[9] And you may recall the suggestion that bison (and perhaps even cowboys) engage in same-sex raunchiness for social reasons. An individual with a flexible sexual appetite can use risqué business to pacify potential enemies, smooth over disputes, and bond with members of their community. Researchers who see same-sex sexual attraction as a pro-social adaptation suggest that as animals evolve to form communities, a noted benefit emerges for individuals who are open to using sex to bond with their peers.

And if sexual fluidity is good for the species, some individuals are bound to end up landing on the same-sex-attraction-only end of the spectrum. Just as some are bound to end up woefully heterosexual and, presumably, no fun at parties.

It's possible that all of the hypotheses I've just outlined are to some degree correct. It's possible that none of them are. And sexual orientation, despite the dogged efforts of many geneticists over the years, can't be codified by simple inheritance; it seems to sometimes be more or less common for members of particular families to lean one way or the other, but it's become clear that whom we love (and whom we want to have sex with) has to do with a complex set of environmental, genetic, and cultural factors. Regardless, there's another, more obvious reason that being queer has stayed in our DNA: gay people have been around for our entire evolutionary history, and some of them have had babies. Some of them may even have had babies with each other! It's important to remember, when we talk about the history of our species, that what we *want* to do is often different from what we *do*.

A look back at ancient Greek culture, for instance, makes clear that the normalization of gay male romantic love did nothing to stop people from breeding. In fact, the Spartan assumption that you'd have much deeper feelings for your male pals (whom you spent all your time with, due to military service, and may or may not have had sex with as a result) than you would for your female spouse may have actually made it more likely for a man with zero heterosexual attraction to help churn out baby soldiers. In general, the default setting for a heterosexual marriage in ancient Greece writ large was a sense of duty and vague indifference, so there was no reason for a particularly gay man to feel a crisis of self when he looked at his bride and thought "meh."

More recent Western culture obviously stigmatized queer attraction. This meant many people hid their orientation from the world or even from themselves. And yet, crucially, many of those people had babies, because that's just what people were supposed to do, as evidenced by the very assumption of a "paradox" we started this chapter by trying to explain.

In other words, we might not need to scratch our heads too hard to figure out how queer-leaning genes have managed to stick around. And if the animal kingdom is any indication, those genes have been with us for a long, long time.

ANIMALS ON PARADE

We're talking about queer animals (including myself), so you're probably wondering where all the gay penguins are. They are here. I'm sorry for withholding the gay penguins from you for so many pages.

In 2012, Douglas Russell came across something wondrous. The curator of birds at the British Natural History Museum found a one-hundred-year-old document detailing the scandalous sexual activity of Adélie penguins in Antarctica: a document written—and then seemingly abandoned—by Dr. George Murray Levick, an officer on the British Antarctic Expedition of 1910.[10] A century ago, Levick saw young males attempt to mate with dead females. He saw them harass chicks. He saw them *have sex with one another*. Museum officials apparently found this behavior so shocking that they published only small snippets of Levick's extensive notes on Adélies, leaving the bulk of his observations on their sex lives to molder in the museum archives. (In Levick's defense, the naturalist did print his notes in a separate tome for private distribution, which one can imagine resulted in some delightfully lurid back-alley scholarship.)

None of this, it should be said, is surprising. As we will see in later chapters, juvenile males of several species have a truly nasty habit of assaulting anything assaultable. And by 2012, scientists had spent decades recounting various sorts of sexual behavior in the much-studied species of bird. As Russell told reporters back in 2012, Adélies meet for just a few weeks a year to rush through a breeding season, and the urgent nature of the task means some of the freshest males will respond to "inappropriate cues" like deadness (which they interpret as sexual submission).

While the twentieth-century paper may not have shocked twenty-first-century ornithologists, Levick was clearly shaken. He even wrote some of what he saw in code (using the Greek instead of the Roman alphabet, though Russell and his colleagues found the encoding of entries to be scattershot and unpredictable). His observations are detailed, but they feature more anthropomorphizing of the penguins than any attempt at analyzing them.[11] Indeed, Levick seems too eager to dismiss the "depraved" behavior of "hooligans" in the penguin world to ask why juveniles might act thusly. He hilariously concludes, "There seems to be no crime too low for these Penguins."

Levick's published work on the animals makes cheeky reference to all of the actual data he left on the cutting-room floor: "The crimes which they commit are such as to find no place in this book, but it is interesting indeed to note that, when nature intends them to find employment, these birds, like men, degenerate in idleness."

It's perhaps understandable that the violent necrophilia at play shocked Levick's sensibilities. But it's telling that museum officials also took the opportunity to cut out observations of homosexual behavior and general promiscuity. When given the chance to offer humans a glimpse into the natural world, museum staff—and countless other arbiters of natural history through the ages—either consciously or unconsciously projected the monogamous sexual mores of the day onto animals. By dismissing very normal aspects of penguin sexuality as sure outliers not worth mentioning, Levick's superiors created an enduring image of penguins as little gents in tuxedos puttering about in married, heterosexual pairs.

And this didn't just do a disservice to the penguins. As we'll see throughout the book, such projecting does a disservice to all of us.

Now, of course, we know that all sorts of penguins can form downright wholesome gay attachments. In 2004, Central Park Zoo chinstraps Roy and Silo famously hatched and fostered a chick from an adopted egg.[12] Other zoos around the world have noted that male couples will often build nests and try to hatch stones and will happily take on joint fatherhood if given an abandoned egg to raise. We should be careful not to make the same mistake as Levick's editors and project our own idea of what it means to be gay-penguin-married on a bunch of birds; no one has seen them perform this level of domesticity in same-sex partnerships in the wild, so we don't know if it's common in totally natural settings. But it *happens*.

If you take nothing else away from this book, know that any kind of sex or sexual desire that you could possibly imagine—and likely even a few types you can't—has *happened* somewhere, at some point, at least once, if not over and over again.

With those British museum officials and their ilk lumping sexual fluidity into the same category as ejaculating into a frozen corpse, it's no surprise that homosexual behavior in the animal kingdom often feels like a newfound phenomenon. It's really, really not. It's as old as the rugged American plains and eons older still.

CHAFING A FEW COCKS

I don't expect you to be particularly compelled to feel one way or the other about insects exhibiting homosexual behavior. With a few notable exceptions in the world of social critters like bees and ants, they're some of the few animals on the planet we generally have no interest in anthropomorphizing. The fact that bedbugs are so indiscriminate in the targets of their "traumatic insemination" (that's where the sperm has to get into the abdominal cavity by way of a stab wound) as to often be "gay" is unlikely to change any opinions on the validity of same-sex relationships or make any queer person feel more secure in their orientation. There are many "bisexual" insects in the animal kingdom, and their same-sex interactions generally seem to boil down to ineptitude.

So, gay bugs are far from glamorous. Nonetheless they have an important place in the history of our understanding of sexuality.

In 1896, Henri Gadeau de Kerville—one of France's leading entomologists—published what should have been a positively pedestrian account on *Melolontha melolontha*, aka the common cockchafer. (At the time of writing, Wikipedia takes great pains to note that the beetle's name refers to the "late 17th century usage of 'cock,' in the sense of expressing size or vigor" and that "chafer" was a generic name for a type of plant-chewing beetle, stemming from a word of Old English origin meaning "gnawer."[13] But I give you permission to laugh at what sounds like a very pointed old-timey insult.)

See, Gadeau de Kerville had spotted some male-on-male action between members of this species, but that was nothing new. Entomologists had been recording (and making excuses for) such couplings for decades. Things would generally start with the scientist in question incorrectly assuming that one half of the pairing was hermaphroditic and had female genitalia despite displaying male antennae. Or they might conclude that the penetrating male had been blinded by lust and had forced itself upon the obviously unwilling, obviously passive receiving male partner.

But Gadeau de Kerville took the stance that some amount of "pederasty" (a term used to refer to the relationships between grown men and teenage boys in ancient Greece) was natural in wild beetles and sometimes happened due to mutual desire for it—and he included a detailed illustration of the gay bugs in question going at it.

This resulted in frankly huge quantities of kerfuffle. On the one hand, Gadeau de Kerville's position seemingly legitimized the existence of a natural inclination toward homosexuality; on the other hand, his colleagues rushed to prove that beetles actually only did gay things when they were, like, really tired and confused. But modern research has shown us that same-sex sexual attraction and behavior persists in animals much more interesting than a common cockchafer.

FLYING SNUGGLE PUPS

Bats get around; scientists have documented same-sex sexual behavior in at least twenty-two species.[14] The scenarios under which this occurs (and the acts that take place) differ from one type of bat to another: some may only demonstrate male-on-male mounting in captivity, for example, where a lack of available females may be to blame. But for some bats, same-sex behavior is a common occurrence in the wild. *Myotis lucifugus*, for example, does spend part of the year mating in what you might consider a traditional fashion. But these little brown bats also have a "passive" mating phase, wherein males continue to mount partners who are dormant and sleepy due to dropping temperatures. During this period the active males go into an indiscriminate frenzy—and a bisexual one at that. One estimate suggests that more than a third of these couplings are homosexual.[15]

Pteropus pselaphon also deserves a shoutout. Male Bonin flying foxes have been shown to engage in frequent and deliberate genital licking in the wild. During mating season, adults cluster into same-sex roosts for warmth, protection, and, apparently, fellatio. In addition to heterosexual genital licking, researchers have also observed frequent male-on-male tonguing. They suspect it may help keep tightly packed dudes from fighting as the mating season progresses.

GAY AS A GOOSE

Your finance guy will probably say that a black swan event is one too rare to predict or prepare for that will dramatically affect the markets or some other societal system. In real life, many black swan events are gay.

Early Europeans should be forgiven for thinking black swans were a once-in-a-blue-moon kind of fowl. They're native to Australia, and Westerners didn't see them until the 1600s, though they've now been introduced into New Zealand, Japan, China, the United Kingdom, and the United States. They're beautiful: strikingly black with red beaks and a flash of white flight feathers that appear when they take wing. They're also queer.

In a 1981 overview of the species, Lionel Wayne Braithwaite reports that *Cygnus atratus* gets the avian equivalent of gay-married with some frequency in the wild.[16] While plenty of male black swans pair up with females and call it a day, both wild birds and those in captivity sometimes enter into male-male pairings instead. These manly family units, according to Braithwaite, present "a formidable combination." They might form "brief associations" with a female before kicking her out to raise the resulting eggs on their own or simply overtake a nest of eggs already produced by another male-female couple.

Same-sex bird partnerships can even occur between different species. A beloved New Zealander goose named Thomas, according to the 2018 BBC obituary for this local icon, took up with an injured female black swan—dubbed Henrietta—who flew into town in 1990.[17] Some eighteen years later, the BBC reports, Henrietta made the acquaintance of another female black swan, and this third started hanging out with their motley crew. Then the new bird laid eggs. Henrietta's eggs. Henrietta was actually a Henry—had, in fact, always been male, though observers hadn't seen fit to realize this. Thomas, meanwhile, who'd had a nearly two-decade relationship with Henry, was suddenly recognized as a gay goose icon.

Thomas, understandably, was less than thrilled to find himself in a poorly negotiated throuple and was aggressive toward the new lady friend at first. But once the eggs hatched, he took on a parental role and helped care for Henry's sixty-eight children as the years went on.

For the two black swans in the group, at least, this arrangement would have seemed perfectly natural: in addition to hetero- and homosexual pairings, *Cygnus atratus* are known to sometimes form long-term trios where two males care for the eggs of one female (without kicking her out once they're laid). All three birds will participate in precopulatory mating displays, but only one male will mount the female, while the other will parade around the nest protectively. Once hatchlings are a couple of months old, the males will take over caring for the nest so the female can lay more. Writing back in 1981, Braithwaite called this system "stable, frequently and repeatedly successful."

ㅈ

Okay, you say. But but but but, you interject. All the indiscriminate rutting and cooperative egg rearing the animal kingdom could possibly provide does not necessarily say a single thing about human queerness.

This is true! A sexually confused steer is obviously not the same as a gay *Homo sapiens* (though I would love to see a cow at Pride, just 'cause they're fun, and 'cause they have that kind of Cottagecore vibe). Let's even say you ignore the evolutionary evidence that heterosexuality is less a default setting and more a state of mind.

Even so, the fact remains that being not-straight wasn't invented yesterday. We may well be entering—or, in fact, already in—an era of particularly gorgeous levels of openness about sexuality in most countries and cultures. But there was *never* a time or place in history where everyone was simply and stably heterosexual. As a species, we have absolutely been there, and we have absolutely done that.

IT'S ALL GREEK TO ME

You probably already know that ancient Greek men had a lot of gay sex. But homosexuality in ancient Greece was . . . complicated, to say the least.

For starters, it didn't exist. That is, people didn't identify themselves or others, like we do today, as being attracted to any gender or sex specifically. Instead, they drew lines based not on whom one was attracted to but, rather, on whether one took active or passive sexual roles—in modern parlance, between tops and bottoms.[18]

So while lots of high-status Greek men had what would now be considered gay sex, they wouldn't have called it that.[19] Similarly, they wouldn't have identified pederasty, a common sexual practice at the time, as age inappropriate, despite the obvious problems we can see with it today.[20]

How did this work? Sometime between age fifteen and eighteen, boys would be considered mature enough for pursuit by a male mentor. Still, they were encouraged to play hard to get so they could tease out the older man's intentions. That's because these partnerships were meant to be deeply involved and mutually beneficial, so there was a real stigma against just trying to pick up a teen for casual sex. Only within this dynamic, in the time before you were a full-grown man, was taking on a passive sexual role considered totally normal. And for the older partner, there was nothing less conventional about playing an active role with a man than there was about having sex with a woman. If two grown men wanted to have a sexual relationship, however, the one perceived as taking the passive role would have faced ridicule.

Yes, this is befuddling. Even the Greeks seem to have been befuddled at the time: Plato famously argued that oppressing homosexuality was equivalent to barbarianism and despotism—and then later argued that homosexuality was unnatural.[21] And if the fact that being the older, dominant male in a relationship with a teenager was seen as more acceptable than being an adult bottom makes you say, "What in the toxic masculinity is this fresh hell," well, same.

My point here (and I really hope this is obvious) isn't to highlight ancient Greek pederasty as some proof that being gay has always been mainstream. The modern conservative movement's tendency to conflate male homosexuality with pedophilia is baseless and disturbing, and ancient Greece was not a model to emulate.

My point in mentioning it is to give you a glimpse of just how arbitrary our societal sense of right and wrong is and how obvious it is, in hindsight, that we tend to get it twisted. The things we normalize and the language we use *matters*. That's how you end up with a culture in which most men enjoy having sex with other men, no one is gay, and a bunch of teenage boys are potentially being pulled into cycles of sexual abuse. I consider this to be great motivation for making sure we don't give thirtieth-century historians fodder to grimace about our sex lives.

As for Greek lesbians? They were a more cryptic bunch. But the famous poetry of Sappho—among other evidence—tells us they at least existed. One of the more complete poems we have of hers captures the feeling of jealousy and longing the writer felt when seeing her female lover with a man:

> *He's equal with the Gods, that man*
> *Who sits across from you,*
> *Face to face, close enough, to sip*
> *Your voice's sweetness,*
> . . . .

And what excites my mind,
Your laughter, glittering. So,
When I see you, for a moment,
My voice goes,
....

My tongue freezes. Fire,
Delicate fire, in the flesh.
Blind, stunned, the sound
Of thunder, in my ears.
....

Shivering with sweat, cold
Tremors over the skin,
I turn the color of dead grass,
*And I'm an inch from dying.**

Given that the men were all busy having totally-not-gay-sex with teenagers, it is perhaps unsurprising that lesbian pining (and, we can only hope for Sappho's sake, a fair amount of actual sex) flew under the radar.

Ancient Rome had a similarly complex and unsettling take on being queer, with the passive role seen as acceptable only for enslaved people. Even so, given the intense sanctity of male friendships in both Greek and Roman cultures, perhaps gay men interested in age-appropriate relationships without horrific power imbalances were able to go mostly incognito.

---

* I should note that, this being an ancient poem that no longer exists in its entirety, there are many translations of it, and I have consciously selected Anthony Kline's 2005 interpretation for having a closing line that could most easily be mistaken for a Mitski lyric.

That changed in the sixth century, by which point Christianity had triumphed across western Europe. Justinian the (not so) Great made it a policy to castrate or even execute anyone found guilty of homosexuality in the name of avoiding the wrath of God, or something.

Ancient Greece and Rome were far from egalitarian paradises for queer people. Still, their stance on same-sex attraction—which is kind of mind-boggling to comprehend using a modern framework—shows us just how much a culture's arbitrary practices can shape what's "normal" and what isn't.

Today, it's a common enough refrain, for example, that youths are out of control with their terms and conditions for describing their sexual identities. Who can keep track of all those letters, how many genders are we expected to assign new stereotypes to, back in my day men were men and women were miserable just as nature intended, etc. But even when mainstream society lacked any language at all to codify differences in sexual attraction or used language that varied wildly from what we use today, people always had sex that wasn't straight and procreative.

PAIRED EATING AND CUT SLEEVES

Once upon a time, a beautiful young man fell asleep on top of the long, draped sleeve of his lover's robe. Yet the trapped man couldn't just lie there: Ai was the emperor of China and had duties to attend to. But, loath to interrupt Dong Xian's peaceful slumber, Ai cut his sleeve off to avoid disturbing his lover.[22] Emperor Ai, who ruled the Han dynasty from 7 to 1 BC, had a wife. But his relationship with Dong was not concealed. Dong held high office, and his family and household members received financial support from the emperor.

Dong and Ai had a tragic end. The childless Ai tried to pass his throne to Dong when he died suddenly. But members of the royal family acted quickly to take power. Dong is said to have committed suicide soon after.

Homophobia probably wasn't the reason the ruling family refused to name Dong emperor. Instead, it was a simple dynastic power play. In fact, according to scholar Bret Hinsch, Ai was the *tenth Han emperor in a row* to pluck a male favorite from his courtiers, and courtly male love was treated as a beautiful thing. Much less was written about lesbian encounters in ancient China, but some experts think notes about "paired eating" between ladies of the court refer to cunnilingus.

This acceptance got rocky during the Song dynasty (AD 960), when an Indian Buddhist text condemning homosexuality became popular reading. And in the 1200s, Genghis Khan outlawed sodomy. Still, in the sixteenth century, more than one Portuguese explorer bemoaned the prevalence of such proclivities in China. Clearly, Khan's smackdown couldn't undo centuries of pro-bisexual sentiment.

PRIDE AND PREJUDICE

The early 1800s was not a good time to be gay in England. In fact, starting in 1533, gay men faced the death penalty in all of Britain.[23] Not until the 1800s was this sentence finally softened to imprisonment—a threat that spread to the countries Britain had colonized, where in some cases it persists today. Even in

1885, the Criminal Law Amendment Act clarified that any male homosexual act in Great Britain, even undertaken in private, was punishable, and the law's wording was so vague that it became known as the "Blackmailer's Charter." For centuries in England, homosexuality was, at best, the kind of thing you had to keep on the down low to avoid trouble.

Case in point: Anne Lister, portrayed in the TV show *Gentleman Jack* and oft called "the first modern lesbian," wrote about her numerous sexual and romantic exploits in code.[24] Even so, it's quite impressive (and telling) that Lister's sexuality could have been considered a secret *at all*, given that she essentially wooed and married a fellow heiress. "She is to give me a ring & I her one in token of our union," Lister wrote of Ann Walker in 1834. While there were no avenues for them to be legally married at the time, Lister's diaries and letters suggest that they saw a joint attendance of Easter Communion at Holy Trinity Church in York as cementing their partnership in the eyes of the Lord.

It's hard to know how much of Lister's "secret" sexuality came down to rich people just letting other rich people do their own thing as long as they were somewhat discrete about it. But her ability to hide in plain sight was also no doubt due to the country's general disregard of lesbianism. The laws on same-sex activity made no mention of women because at least some men in power thought gay women were extremely rare. Even when the British Parliament considered adding antilesbian language to existing laws in 1921, the Houses of Commons and Lords decided that, actually, putting female gayness on the books would only make more women aware of lesbianism, which might encourage them to explore it (which suggests the members of Parliament had *just swell* relationships with their wives and daughters).

Despite these centuries of death penalties and condemnations, there's reason to believe that not all everyday citizens took a negative view of homosexuality. In 2020, historians from Oxford University discovered a diary from 1810 in which Matthew Tomlinson, a middle-class farmer, recounted his discussions about and evolving thoughts on gay men.[25] Tomlinson was mulling over a scandalous court case involving a naval surgeon, who was sentenced to hang when found to be engaging in gay acts. Tomlinson granted that if homosexuality was a choice and therefore a sin, some punishment like castration might make sense to save mortal souls. But he also laid out the facts he'd collected from conversations with others and seemed to conclude that homosexuality could develop early in life as opposed to resulting from a fall from grace. If gay men had an "inclination" and "propensity" to be so, he wrote, then being gay "must then be considered natural." If the divine creator created gay people, he wondered, then what right did the government have to punish them for it or try to change them?

Tomlinson was just one man, but his genuine curiosity on the subject—together with the fact that he seems to have discussed it with other members of his social sphere—has made historians curious about how everyday folk might have perceived queerness.

This is a lesson we should keep in mind when we reflect on all eras of history—not just the days of Jane Austen's England. Even in the many, many instances where a government has opposed homosexuality, it's unlikely that every straight person living under that government agreed.

A BOSTON-STYLE MARRIAGE

Here's the thing: Victorian women were encouraged to form close female friendships, and in some circles, those pals regularly held hands with, snuggled down on, and kissed one another. Lesbianism just wasn't *supposed to be a thing*, you see; in fact, women were generally expected to keep themselves pure of all sexual vice. Yes, many of these touchy-feely friendships were no doubt truly platonic; there's a reason queer women have a million jokes about not knowing when someone is flirting with them, okay? Still, even in the late 1800s, people started to get wise to the idea that some of these women . . . were probably having sex.

In 1886, Henry James published *The Bostonians* and inspired the term "Boston marriage"—like the characters in his book, women in these northeastern arrangements eschewed the company of men in favor of shacking up together.[26] Some even exchanged rings and vows. While some Bostonians may indeed have just been very good friends, James drew his inspiration from a couple who were almost undeniably in love. Annie Adams Fields was married to publisher James T. Fields for more than twenty years but paired up with poet Sarah Orne Jewett upon his death. They spent the next three decades living, traveling, and hosting salons as a couple. If you have any inclination to doubt their romantic love, take a peek at this poem Orne Jewett is thought to have written for Fields or another gal pal:

> *Do you remember, darling*
> *A year ago today*
> *When we gave ourselves to each other*

Before you went away
At the end of that pleasant summer weather
Which we had spent by the sea together?
. . . .

How little we knew, my darling
All that the year would bring!
Did I think of the wretched mornings
When I should kiss my ring
And long with all my heart to see
The girl who gave the ring to me?
. . . .

We have not been sorry, darling
We have loved each other so—
We will not take back the promises
We made a year ago—
. . . .

And so again, my darling,
I give myself to you,
With graver thought than a year ago
With love that is deep and true.

In modern times, some straight women who've chosen to live as long-term platonic partners with other straight women have appropriated the phrase "Boston marriage" for their own.[27] I think this is probably fine. Surely some of the independent ladies who were side-eyed for their housing choices in Victorian New England were not, in fact, sexually involved with one another and were just, in fact, tired of men trying to claim their inheritances and ruin their lives. But please respect the lesbians who walked so you could run into signing a mortgage with your "heterosexual soulmate."

WELCOME TO WEIMAR

Just before the Nazis rose to power, Germany had a thriving queer scene.[28] Sex between men was illegal—had been since 1871—but in the early twentieth century, gay culture thrived anyway. Berlin had something like one hundred gay and lesbian clubs and cafés in the 1920s. And in 1919 the city saw the founding of the Institute for Sexual Science (Institut für Sexualwissenschaft), the world's first institution devoted to studying sex. The center's founder, Magnus Hirschfeld, a German Jew, was an early champion of queer and trans rights.[29] Still, he wasn't universally beloved even within the community he fought for; some gay men took issue with Hirschfeld's support of effeminate men, trans people, and feminists, and he faced anti-Semitism that equated his liberalism with anti-Aryan sentiment.

But Berlin *was* an incredibly queer city. Some historians say that a vote to decriminalize homosexuality was in the cards in 1929, before a stock market crash sent the country into turmoil and paved the way for Adolf Hitler's rise to power.[30] Unsurprisingly, Nazis destroyed the institute, Hirschfeld fled the country, and many of the queer individuals who thrived during the days of the Weimar Republic were imprisoned or killed during the Holocaust. Things don't always happen as linearly as we'd like them to. Don't ever let anyone tell you that being queer is a recent invention.

WHAT DOES THE FUTURE HOLD FOR HETEROSEXUALITY?

Let's review what we've learned so far. Straight sex has never really been the status quo that many of us were taught to believe. Yes, it is what the majority of humans seem to be into, which probably has to do with the way our species evolved and what strategies our distant animal ancestors used to diversify their genes. But still, it's probably *never* been a universally applicable preference.

In fact, it's possible that our heritage is one of sexual ambidextrousness and that we simply picked up heterosexuality along the way. Nature has always made a place for queerness. It's a healthy part of a balanced breakfast, evolutionarily speaking.

But it's understandable if you sometimes feel like the world is getting gayer by the minute.

A 2020 Gallup poll found that *one in six* adult members of Generation Z (people aged eighteen to twenty-three in 2020) in the United States identify as something other than heterosexual.[31] That's almost 16 percent and comes just five years after millenials were dubbed the gayest generation ever for just a 7 percent rate of LGBTQ-ism.[32] In the 2020 survey, some 5.6 percent of all US adults self-identified as members of the alphabet mafia, which was up just a touch more than a full percentage point from the 2017 poll. In 2020 more than one in ten of all adult members of Gen Z in America identified as bisexual.

That's a lot. That's nontrivial. And given that bi people are less likely to be out or to outwardly present as queer—this is especially true for men—one has to wonder just how many potential bisexuals hesitated before claiming heterosexuality when Gallup asked.

It is, however, impossible to know whether more people are queer than they were fifty years ago or whether queer people now are just more likely to recognize and speak to their orientation. Given the rich tapestry of sexual norms that have existed throughout our species' history, I'm going to put in an enthusiastic vote for the latter possibility. Probably—whether they could say it or act like it or not—a good part of the species has always been pretty queer.

Of course, queerness isn't just about sexual orientation. In modern parlance, it's also about biological sex and gender identity. But don't worry: we can bust up that binary too.

Chapter Three

JUST HOW MANY SEXES ARE THERE?

....

In which slime molds, ligers, and songbirds help us
understand the complexity of biological sex.

....

IF *FINDING NEMO* HAD BEEN SCIENTIFICALLY ACCURATE, IT WOULD HAVE
started with a sex change. All clownfish are born male, but some
of them are lucky enough to turn into females. I'm not being
misandrist for the heck of it, folks: it's good to be a lady in the
clownfish world, because girls run the show. In fact, a 2019 study
from the University of Illinois found that when two male clownfish are put together in isolation, they'll fight—and the *winner*
will develop into a female so the pair can start mating.[1]

Generally, a female and a male will shack up together inside a
stinging anemone for protection—they're immune to its stings, as
you might recall from the aforementioned Disney film. If other

females show up, the biggest lady will fight the rest off. Any additional males who try to join the crew will just hang around without trying to mate. They're waiting for the opportunity that arises in the case of the alpha female's death (which isn't an unlikely prospect, given that she does all the work of defending the anemone from outside predators). Because if there's only one female in town, when she dies, her mate will change its sex and shack up with the next-largest guy around.

According to that 2019 study, the process starts in the brain. As soon as a male establishes dominance over the rest of the crew, he starts acting female (read: like a boss). This is reflected, as weeks and months go by, in the preoptic area of the brain—the region that regulates sexual organs. Marked differences in this region correspond with sex: by six months down the road, a male who's taken on the top leadership role will have a brain indistinguishable from a fertile female's, even though the fish can take months or years to develop the corresponding sex organs.

So, yes. If *Finding Nemo* were biologically accurate, Merlin would have immediately ditched his timid nature and started to hashtag lean into being a hashtag lady boss, and eventually completed a full sex change. Unfortunately, unless a rogue male clownfish happened to wander by the anemone homestead before Nemo reached maturity, the laws of nature would have dictated that the little guy would not only maintain his maleness but mate with his now female dad. All in all, it seems reasonable that Disney decided to take some poetic license on this one. Don't worry. Clownfish are far from the only organisms that make the sexual binary get red in the face.

BITS AND BLOBS

In 2019, I came across something suspicious on my Twitter feed. It was a Reuters article about the so-called mysterious new resident of a Parisian zoo: a blob with 720 sexes that displayed both fungal and animal characteristics. "I was confused," I wrote in *Popular Science* at the time, "because the blob was not at all mysterious. The photo clearly showed a slime mold. In fact, the photo caption even said it was *Physarum polycephalum*. That's not even a new slime mold. It's actually, like, the most pedestrian of all the slime molds."[2] As a dogged spokesperson for sexually complicated blobs of all sorts, I felt particularly offended.

I will spare you a full treatise on the slime mold. To briefly summarize, they're not actually molds but protists, and they don't really have anything in common with other members of their kingdom (protists, it should be said, generally have a real "put this one over there with the other weird stuff" vibe). Slime molds are technically single-celled organisms, but they can mate and form synchronized colonies called plasmodium, where all of their nuclei chill in one giant glob of harmonious intracellular matrix that will scramble back together if sliced apart. They can move, hunt, and forage in sync. They don't have central nervous systems, but they seem capable of learning and adapting their behavior to suit their environments.

They also have hundreds of sexes.

A sex, at its core, is just a category that determines how you contribute to the reproductive process of your species—in the case of an organism as simple as a slime mold, it refers to what type of cells you make that are designed to combine with other

cells to form a zygote. When slime molds release their reproductive spores into the air, they send out sex cells that feature
three variable genes. But having three sexes would be too simple
for these slimy lads: each of those genes can come in a bunch of
slightly different varieties, and each mature individual carries
two copies of whichever versions of the three genes they've got.
Slime molds don't have an "opposite sex"; each can combine
with some seven-hundred-plus sexes other than its own. The
mind boggles!

And slime molds are nothing compared to *Schizophyllum commune*, a fungus that invades rotten wood. These fungi have so
many points of potential variation in the DNA that controls their
gametes that they come in more than twenty-three thousand different sexes.[3]

FREE BIRDS

Chickens are simple creatures, no? No. They're not. Not when it
comes to sexual development. Little did you know that every hen
you see exists in a liminal state. She carries within her a secret:
the ability to become male.

See, chickens do all their prolific egg laying thanks to just one
functional ovary (the left one, to be precise). As is common in the
animal kingdom, during a chicken's embryonic development, its
future sex organs contain a world of possibilities, with the ability
to grow into either type of gonad. For whatever reason, in hens
typically only one of two gonads (lefty) actually makes the transition into an ovary before the birds hatch. The right one tends
to stay ambiguous—a sort of choose-your-own-adventure gonad
that can be activated later in life.

If some kind of illness or injury damages that trusty left ovary, the right one wakes up—but not necessarily as an egg-producing backup. Instead, the drop in estrogen due to the busted gonad and a resulting rise in androgen can trigger the development of the ambiguous right gonad into an ovotestis. That's an intersex organ, but in this case, it acts like testes. That means the bird's androgen levels shoot up even higher.[4]

Throughout this process a hen will start looking and acting like a rooster; she'll grow wattles beneath her chin and a cockscomb on her wee head and begin to strut and crow and generally make a fuss. She'll also stop laying eggs.

It seems like this phenomenon is almost exclusively seen in chickens that started out female and is unlikely to lead to actual sperm production and fatherhood. But some chicken experts have heard tell of former hens actually producing sperm or of roosters turning into hens and laying eggs.[5] For some reason, these strange happenings have not been researched thoroughly, so the rate at which chickens perform this switch—and the reproductive implications—remain murky.

LIVE, LAUGH, PSEUDOPENIS

I wasn't sure whether to put the hyenas with the sexually fluid animals or the animals with fluid sexuality. I'm putting them here because we all know you're only going to care about their pseudopenises, which is more of a "sexual development is more complex than you've been told" point than a "damn, these bitches gay, good for them" point. I think. I don't know, I'm not your dad, you can disagree with me if you want. Writing a book is hard even when you're not trying to arbitrarily delineate a series of topical themes from the pile of mud and bones and confusing genitals that is the natural world.

We have reason to believe that humans long thought hyenas to be hermaphrodites, meaning that they had physical features associated with both male and female bodies—or even that they were capable of swapping sex at will. Two of Aesop's Fables from the sixth century BC use hyenas' supposed annual sex changes as a plot point. In one, a disrespectful male is warned by the female it pesters that next year she might be in a position to impose the same assault on him. In another, a male fox rejects a would-be hyena lover not because they're two different species but on the grounds that she might up and switch sexes on him.[6] In the first century, both Ovid and Pliny the Elder commented on the beast's bizarre nature. "We might marvel at how the hyena changes function," Ovid wrote in *The Metamorphoses*, "and a moment ago a female, taken from behind by a male, is now a male." Pliny, in addition to claiming that hyenas were wolf-dog hybrids (not true), were incapable of bending their necks (baffling), could imitate the human voice (sort of, but not in the way he meant), would call you by your name to lead you to your death (lol), and could strike other animals dumb or paralyze them (cool but wrong), also repeated Aesop's notion of an annual sex-change jubilee. Aristotle reportedly disagreed with this take, so it apparently wasn't a universally accepted falsehood even in the ancient world. Still, it wasn't until the late 1800s that English anatomist Morrison Watson would formally combat this assumption, and the myth persists among some laypeople to this day.

In a 1981 paper, famed biologist Stephen Jay Gould called the accusation of hermaphroditism among hyenas "the unkindest cut of all."[7] Gould found this unkinder, apparently, than even reports that they preferred putrefied flesh and would gorge themselves until all they could do was poop and vomit simultaneously. This says a lot about biology in the 1980s, I guess.

Okay, but why do hyenas inspire this sexual confusion? It's because the females have penises.[8] Sort of. There are absolutely biologists who would set me and everything I own on fire for calling the appendage attached to female hyenas a "penis," but I am not here for pedantry. These ladies have dicks. Johnsons. Peckers. Schlongs. Willies. Wieners. I could go on.

When individuals who are functionally "female"—in terms of the part they play in reproduction—have protrusions resembling penises, those organs are generally referred to as pseudopenises. This is because they only *look* like a fully functional penis in the reproductive sense; a human might choose to use an enlarged clitoris for penetrative sex, and you are more than welcome, of course, to rechristen one as a penis in all the ways that matter to you, but it's not going to produce ejaculate, and you're not going to pee out of it. But in the case of the spotted hyena, while the thing that looks like a penis may not provide an escape route for sperm, it's *absolutely* crucial to both the act of mating and the process of reproduction (plus, for what it's worth, they absolutely *do* pee through it).

See, *Crocuta crocuta* doesn't have an external vaginal opening. It has to urinate, have sex, and give birth through its massive clitoris, which when erect is almost indistinguishable from a male hyena's penis in size and form. The female even has what looks like a set of testicles, though it's really just a rather plump pair of labia fused together. Females walk around with proud erections on the regular, as sporting one is actually a sign of *submission* to the alpha lady of the pack.[9] And yes, unsurprisingly, female hyenas, who are also bigger and more aggressive than their male counterparts, call all the shots.

Sex, as you might imagine, is complicated for hyenas. When the pseudopenis is flaccid and the female is cooperative, it is possible for a male to angle his shaft into such a position that it can slide into her own. One cool feature of the all-purpose sex tube is that it's likely that urinating after sex flushes all the sperm out—it's all going through the same opening, after all— which gives the female *Crocuta* even more power over the males of her species.

Birth, it should go without saying, is a horror. Hyena cubs come out fully armed (with teeth) and ready to rumble, which means their mothers push three pounds of squirming muscle out of their flaccid members. The process often ends in death for mother, cub, or both when a female delivers her first baby, since the pseudopenis has not yet been stretched by an older sibling and the route out of the uterus is awkwardly long and bendy.

No one is quite sure why *Crocuta crocuta* evolved this way. The setup obviously has its downsides. But it's possible that these pseudopenises are an extreme result of a war of the sexes, where females were more likely to survive and have offspring if they had the ability to dominate the males of the species and control sexual encounters. The whole giving-birth-through-a-narrow-tube thing is just a reminder that evolution does not take any personal requests under consideration and that many instances of natural selection turn into a real sort of "Monkey's Paw" situation.

SINGING A DIFFERENT TUNE

If sex is about what cells and parts your body contributes to the equation of reproduction, then *Zonotrichia albicollis* is a bird with four sexes. After some twenty-five years of field study on white-throated sparrows, researchers led by Elaina M. Tuttle published reams of genomic data in *Current Biology* in 2016.[10] Their findings were, on the one hand, pretty straightforward: they identified certain genes associated with the two different color schemes the birds tend to come in, white and tan. But instead of having just a handful of random snippets of DNA that code for color, these two morphs are divided by a "supergene": an array of more than eleven hundred genes that are totally inverted in one group of birds versus another and always came as a complete set.[11] When you look at the sex chromosomes of a white bird versus a tan one, it's as if someone sliced out a segment from one of them, put that thing down, flipped it, and reversed it.

That one-eighty has big implications borne out in the way the birds mate. To reproduce successfully, it seems that *Zonotrichia albicollis* has to find not only a member of the opposite sex but a member of the opposite sex who *also* has the opposite supergene setup. This is especially fascinating given that the morphs tend to have different social characteristics: white birds sing tunefully, act promiscuously and aggressively, and have little care for their offspring; tan birds tend to be bad singers and protective, monogamous mates who fuss over their young.

Tuttle and her colleagues demonstrated that the mutations on the affected chromosomes were accumulating faster than genetic mutations elsewhere in the birds' DNA, essentially doubling down on the need for birds to mate with members of the opposite morph. That seems an awful lot like new sexes are developing—creating a world where *Zonotrichia albicollis* has four sexes, and each can only mate with one of the other three.

That might sound wild, but that's pretty much how scientists think XX and XY chromosomes evolved in humans too. And so-called biological sex? Even after millions of years of evolution, it's still pretty complicated.

HUMANS

We accept, as a society, that while around 2,999 out of 3,000 babies will emerge without teeth, that lucky 3,000th player will come out with a few discomfiting chompers. A newborn with pearly whites (just a few, mind you) is undeniably different from the *average* infant, but this quirk doesn't rock our understanding of babyhood or teeth. It's just a thing that happens sometimes and a reminder that what is now so typical as to be assumed normal for our species was likely not always thus. Some cultures see so-called natal teeth as good or bad luck, but the same can be said for more common variances, like certain eye or hair colors.[12]

We're still working on having the same grasp of differences in sex development, which one academic review in 2000 estimated could affect as many as 1 to 2 percent of all births.[13] Other estimates clock in lower—making such differences from the standard binary sex as uncommon as one in five thousand births or so—but they are not exceptionally rare.[14]

There are dozens of ways in which the story of sex you've likely always been told—that Mommy gave you one of her two X chromosomes and Daddy gave you either an X or a Y, which determined whether you grew a penis or a vagina in the womb—can actually play out. Sometimes an XY embryo can lose its Y along the way (think of it like a biological file corruption), leading to a mosaic of cells in various states of sexual being, which can result in any number of morphological changes to the standard template. Sometimes extra Xs tag along for the ride, leading to a baby with XXY, XXXY, or even XXXXY chromosomes, who might develop "normally" or might present with infertility and a lack of "masculine" body traits like facial hair growth. Sometimes one of your two X chromosomes has accidentally towed a bit of Y along with it. Sometimes a single X shows up to the party, and that's all you get.

Even when the whole XY or XX tango goes smoothly, it's not horribly rare for sex differences to arise. Sometimes the part of the fetal body that would, in most people, develop into one set of genitalia or another just doesn't quite get there for reasons unknown. You might be somewhat or completely unable to utilize the hormones associated with your assigned sex at the cellular level, which can lead your growing body to take shape in unexpected ways. Your organs may simply struggle to produce any sex hormones at all.

All of these varieties in sexual development (and many more) have been documented in multiple patients, and only some of them put a person's health at risk. Point being, the development of a "biological sex" is already more complicated than many of us are told in school. And even individuals whose chromosomes match their reproductive organs can run into some trouble with the false binary.

I recently found out that my own body is staging a minor sexual revolution: I produce "too much" of a "male" hormone for things to work the way they're "supposed" to. Polycystic ovary syndrome (PCOS) isn't an intersex condition, per say, but it does give one something to think about.

In 6 to 12 percent of assigned-female-at-birth people of reproductive age in the United States, the ovaries produce more follicles—the spots where eggs are supposed to pop out once a month or so—than is typical. Those follicles produce androgen, a hormone associated with masculine physical characteristics, which leads some bodies with PCOS to carry more of the chemical than is typical. PCOS is best known as a cause of infertility, since the hormonal imbalance and ovarian irregularities can keep a patient from ovulating at a regular clip. But it can also lead to a laundry list of poorly understood symptoms from fatigue to inflammation; unsurprisingly, the issues that get the most attention are a higher predisposition to diabetes, increased abdominal fat, and "male" hair growth patterns (loss on the head and an increase everywhere else).

Getting diagnosed was frustrating, mainly due to the lack of research to back up common medical consensus on treatment (note: don't let a doctor tell you to go on a low-carb diet, which is based on outdated bullshit and will only make you want to murder everyone). But it also made me wonder: Was this why I had gradually come to see myself as a "woman" in the same way people who go to mass on Easter and Christmas are "Catholic"? The short answer is "who knows"—lots of men, women, and nonbinary people have higher or lower androgen or testosterone without a corresponding shift in identity. Hormones do not a sex, gender, or orientation make. Some small studies have suggested that having PCOS makes you more likely to transition to male, but other investigations have failed to support the connection.[15] Still, it says a lot that researchers are asking.

Some people with PCOS do consider themselves to be intersex, because of either the intensity of their symptoms, the simultaneous diagnosis of an indisputably intersex condition, or simply personal preference. I don't, because my PCOS has not been medicalized, stigmatized, or abnormalized in the same way or to the same extent as diagnoses that involve chromosomal or genital changes, and my hormonal "imbalance" is minor enough that no medical professional would question my sex.

But it does solidify my belief that treating human sex as a strict binary is, at best, a waste of time. If up to 20 percent of the global female population has "abnormal" sex hormones due to PCOS alone, just how normal can the "normal" ones really be?[16]

WE KNOW LESS THAN WE THINK

The medical establishment struggles to define "normal" male and female bodies all the time; as recently as 1999, the *Journal of Pediatrics* published a completely straight-faced study asking whether babies with unusually small penises might be better raised as girls.[17] (Thankfully, even those study authors concluded that, at least given the lack of studies on long-term outcomes for such a procedure, it was probably best to wait and see.) Similar arguments have been made, by the way, about babies with particularly large clitorises—though generally the thinking is not to raise them as boys but to cut off a body part full of nerve endings just because it makes grown-ups uncomfy.[18] As part of the "pro" column, the 1999 study authors mentioned the work of psychologist John Money. That citation says a lot.

In the late 1960s, seven-month-old twin boys named Bruce and Brian Reimer were booked for circumcisions in an attempt to fix an issue where their foreskins got in the way of urination. Doctors used electrocauterization on Bruce—literally attempting to burn away part of his foreskin, which is not how most people perform the delicate surgery—and damaged much of his penis beyond all hope of repair. Understandably distraught about how this might affect their child's life, the Reimers eventually caught wind of the work of psychologist John Money and sought him out.[19]

Money's view on gender was, in some ways, pretty progressive for the time: he argued that gender identity was just about entirely based on social conditioning and that people of any chromosomal disposition would bend to the pressures placed upon them to act "masculine" or "feminine."

But while our modern-day version of "gender is a societal construct" leads some mental health professionals to encourage children to explore their identities with a kind eye toward fluidity, Money did the exact opposite. With the Reimer twins providing what he saw as a perfect experimental set, he convinced their parents to raise Bruce as a girl.[20] They changed his name to Brenda, had him undergo vaginoplasty, used hormonal therapy to give him the desired secondary sex characteristics, and, until the boy's teen years, allowed Money to subject him and his brother to "gender affirming" therapy—which, according to later reports from the twins, included such sexual abuses as the psychologist forcing them to undress in his presence and simulate heterosexual intercourse with one another.[21]

You'll notice that I'm using male pronouns throughout, and that's not a mistake. Around age fourteen, "Brenda" became depressed and pushed back against both his assigned gender identity and the treatment he received from Money. When his parents told him the truth, he shed all the trappings of femininity, chose the name David for himself, and set out to undo the surgical, chemical, and psychiatric interventions that had colored his childhood. David took his own life in 2004.[22]

Meanwhile, Money and his colleagues imposed similar surgeries on countless other infants. That's the sort of work that was still getting him cited in journals in 1999. Part of Money's gender-is-what-you-make-of-it schtick was the notion that a child who was born noticeably intersex—that is, with external genitalia that didn't quite fit the standard mold for either male or female—should be neither treated according to any medical conclusion on which sex they "really" were nor given the time and space to figure it out for themself. Instead, Money argued that parents and doctors should immediately pick whichever sex it would be easier to make them *appear*. If, as Money posited, a child raised as a girl would happily consider themself a girl, then it followed that the best way to ensure the happiness of a child born more ambiguously was to lop off anything that didn't look like a vagina and clitoris, put them in a pink blanket in the nursery, and call it a day—even if they were born with testes in addition to, or perhaps instead of, ovaries and a uterus. Starting in the 1960s, this logic even saw infants put under the knife to right "wrongs" like especially large clitorises or small penises, as the thinking was that no one could possibly function in society with *any* ambiguity as to which role they might play in penetrative sex.[23]

Even if you were born with an unambiguous sex and have a gender identity that matches it, this thinking has negatively affected you. Know this: just as a society that thrives on white supremacy makes even white bodies suffer when their hair texture, facial features, or body shape don't match the supremacist ideal, a society that imposes a strict sexual and gender binary is not unkind only to intersex, fluid, or trans people. Drawing these lines in the sand makes it harder for *anyone* to accept their body and what they'd like to do with it. Whether you're medicalized or made to feel less-than for having external organs a little bigger or smaller than average, or for having a "boyish" figure or a "delicate" face, or for having "too much" of one hormone or "too little" of another, understand that working to liberate those oppressed by the sex and gender binaries will make *your* life better too. I would hope that's not the only reason you'll fight for the bodily autonomy of others, but, hey, just in case—consider it some food for thought.

AN ABUNDANCE OF GENDERS

Sex is complicated, right? And gender is also complicated. But it would be fair to assume that starting from the baseline of male and female—two sexes meant to go together—is the norm, because that's what most of us are taught. Fortunately, loads of cultures work with an entirely different playing field.

In Albania, a dwindling population of people assigned female at birth choose to be *burrnesha*, or sworn virgins.[24] A fifteenth-century code gave them the right to dress, behave, work, and inherit property as men could if they pledged celibacy.

Before the British colonized swaths of South Asia in the nineteenth and twentieth centuries, a culture of third-gender people called *hijras* flourished (and was at times treated as mystical in

Hindu communities) for hundreds of years. *Hijras* are assigned male at birth but present characteristics generally associated with femininity.[25] Only in the twenty-first century, after more than one hundred years of discrimination encouraged by British colonial law, did India, Nepal, and Bangladesh move to formally recognize a third gender category.

Tracing back to the precolonial era in Uganda, the Lango people recognize the existence of *mudoko dako*, who are assigned male at birth but live as women in adulthood. A member of this group can marry a man without judgment or punishment.

Native Hawaiian and Tahitian cultures historically understood the validity of *māhū*, who are assigned male at birth but identify as something in between male and female. Unfortunately, they found their position as respected members of society put in jeopardy as colonizers imposed outside gender norms on the islands. That colonial influence—including from the United States—has continued to negatively affect Hawaiian gender tolerance.[26]

Samoans who are assigned male at birth and live as women, gay men, or a nonbinary third gender are known as *fa'afafine*, while people assigned female at birth who live as men or lesbian women or have a nonbinary identity are called *fa'afatama*.[27] Some of these individuals also identify with the term "transgender," but others don't; Samoan culture already gives them a name that doesn't imply they've changed something about themselves.

Among the indigenous nations of North America, having words for and an understanding of more than two gender identities is so common that, in 1990, a group of tribal representatives voted to adopt the term "two-spirit" as an umbrella to cover them all.[28] This was done not because the myriad identities historically accepted by different tribal nations were one and the same but in recognition of the fact that all were imperiled by US laws and the imposition of European culture.[29]

WHERE DOES GENDER GO FROM HERE?

I'm not trying to take your maleness or femaleness away from you, if you consider yourself a lifelong member of one sex or the other. If you are staunchly and unambiguously male or female down to your DNA, that's okay, man (or lady), and no one is trying to tell you otherwise.

I think, too, that most biologists—even those who espouse the notion that sex is more a spectrum than a binary—would agree that *many*, and perhaps even *most*, humans fall pretty squarely on one end of the rainbow or the other. If that weren't true, it would have been difficult for our binary understanding of sex and gender to crop up and flourish and become overgrown.

But it *is* overgrown; many humans who fit near enough into the slots called male and female to get by don't *entirely* fit there, and even from a purely physiological standpoint, the spectrum is *full*. There are people whose chromosomes are more complex than XX or XY. There are people whose bodies don't produce the hormones that usually go with those genetic scripts. There are people whose bodies don't grow to physically resemble what is typical for what would otherwise be considered their sex. In a world where your sex ultimately boils down to which role you're born to play in the act of reproduction, there are plenty of people who do not have the ability to play either.

The male and female buckets don't have to be fake for humanity to realize that the many people who land outside them (in the middle, just to the right, on the other side of the goddamn room, whatever) are more than just incidental splashes. Sex isn't the only biological distinction that seems incredibly complicated once we take a closer look. I promise we've been here before.

I want you to consider something for a moment. What's the definition of a species? What were you told back in school? Chances are good you learned something to the tune of "a group of animals capable of reproducing with one another." A dog can't make babies with a cat. Dogs and cats are two different species. Okay. That's pretty logical! It's also wrong.

Mules are, of course, an inconvenient exception to the rule above: they're born of the unholy and unnatural union between a horse and a donkey. But wait, you say, this is an easy fix. A species is a group of animals capable of reproducing and *creating fertile offspring* with one another. Well, sorry, but female ligers (that's the perplexing spawn of a male lion and a female tiger, not to be confused with tigons, which are the result of the reverse) can usually make babies with lions. Female tigons can breed with male tigers. If that's not compelling enough for you, consider that our own ancestors hybridized with other species and had off-spring who lived to tell the tale (me, it's me, I'm telling the tale, though actually genetic testing tells me I have a relatively low ratio of Neanderthal DNA, which is a bummer, so I'm holding out hope for a test that reveals me to be full of Denisovan genes or, ideally, something even weirder).

This would be enough to break our long-standing concept of what a species is, but things get more bizarre. Several "species" (cough cough) of the salamander genus *Ambystoma* have been female-only for millions of years.[30] But they don't rely on anything as pedestrian as asexual cloning to keep their good thing going. Instead, these lizard ladies—tied together only by shared mitochondrial DNA, which is the parcel that mothers pass on wholesale to their offspring without any paternal DNA cutting in—mate with males of several species in their genus to collect an assortment of genes. They use some combination of these genes, or none of them at all, when they create their children. Research suggests the lineage has at times gone for millions of years without any outside influence but that mamas can, at any point, choose to shake things up with a random smattering of stolen DNA. A mom might even pass along a combination of up to five different male genomes while leaving her own genes out entirely, save for the mitochondrial DNA that makes this "species" an identifiable family.

This method of reproduction is called kleptogenesis, which is one of the coolest words you'll ever learn. And it results in a "species" defined by only the tiniest shared portion of DNA—and by a tendency to screw around with *other* species.[31]

That definition of species you learned in grade school seems obvious and true because it works *most of the time.* And it's a little mind-blowing—scary, even—to try to come up with an alternative. Multiple scientists in multiple disciplines are sputtering over the conundrum at this very moment, I'm sure. It would be easier to plop down on the ground and say humbug to anything that breaks the simple rule. But that doesn't change the fact that the rule has been broken, and is going to continue to be broken, and is not a universal truth.

Humans love drawing boxes around things. We love creating categories. It's natural: being able to lump X together with Y and distinguish both from *Z* helped our ancestors make quick decisions about which animals, plants, and environmental hazards to chance an encounter with or run away from. But it's become abundantly clear that many of our hastily drawn categories don't reflect the messy reality of nature. That doesn't mean we have to pretend a lion is the same thing as a tiger, and it certainly doesn't mean we expect chimps to start mating with kangaroos. Our framework can grow and change and melt and shift without being obliterated.

Binary sex is much the same: the two categories we came up with aren't inherently *wrong*; they just don't tell the whole story. The lines are blurrier than previously assumed. Sex, like so many aspects of our humanity, is just more complicated than we could possibly have hoped to get a handle on for most of history. That doesn't have to be scary. It doesn't have to be radical. It doesn't have to change the way you see yourself. But if the idea that we might have oversimplified the concept of sex at some point seems antiscientific to you, then I have to point out that you haven't been paying much attention to the way science works at *all*.

Chapter Four

HOW DO WE DO IT?

· ·

In which our cousins the bonobos show us a better
way to live and Puritan teenagers share beds.

· ·

SOMETIME AROUND THE TURN OF THE TWENTY-FIRST CENTURY, I EN-
countered a pop star's interview in some gossip rag or another.
I'm pretty confident this was either Pink, Gwen Stefani, or
Fergie, but I refuse to dedicate any brain space to figuring out
which of the three. Anyway, said songstress recounted her dating
history, and someone—either an adult in my life reading over
my shoulder or a talk show host commenting on her quotes—
proclaimed her a "serial monogamist." The term really struck me.
Having grown up attending evangelical Sunday school, I didn't
have much of a concept of monogamy—I just knew I was supposed to pick one person and stick with them.

Even if I didn't end up saving myself for marriage, there
was an implication that while one premarital partner might be

considered an understandable lapse in purity, I should ultimately endeavor to have as few sexual partners as possible. Indeed, even if I *did* avoid doing the deed until my wedding night, I was told that any dating-adjacent activities I indulged in with someone other than my future husband—kissing, flirting, holding hands, gazing longingly at one another over a box of cheese fries at the bowling alley on black-light night—would serve to dilute the sanctity of the love I could one day offer as a wife (and, perhaps more importantly in the eyes of my youth pastor, just generally make people think I was a big ol' slut).

The idea that one could commit to many people sequentially fascinated me. I suddenly became aware of a reality where one could treat love and sex with some degree of reverence without marrying the first boy they gave a hand job to at the movie theater.

As far as my church was concerned, the idea of serial monogamy—not to mention relationships that existed outside the bounds of monogamy entirely—had emerged recently as the result of society's slip into the muck of moral decay. Shacking up with one person for life and doing so with penis-in-vagina virginity intact was a hallmark of civilized society, I had always been told, and certainly a hallmark of Christian society. *Dating* without the intention of marriage was a new, sinful invention glorified by sitcoms and Nickelodeon shows. We attended sermons and seminars and worship concerts designed to convince us that we should be, like, super punk rock and counterculture by making the radical decision to instead court our future spouses at church events *only*. It is difficult to describe how it feels to be, oh, maybe thirteen years old and realize that your youth pastor dearly hopes you will marry one of the ten male tweens he forces you to spend time with every week (and that he really hopes you'll do so ASAP).

But even in most secular circles, there persists a belief that the way modern dating works—and the way "hookup culture" has emerged around it—represents something new and potentially dangerous. In some ways, that might be true; while data on casual sex among young people shows that many of them enjoy it, many participants (especially women) report mixed or negative feelings around encounters.[1] This may be due to deeply ingrained shame, in some cases. Still, anyone who was at all recently in college knows that there's often a serious lack of communication around sexual boundaries and desires, making one-off or casual arrangements all the more likely to leave someone feeling icky.

Even so, this notion—that romantic relationships have gone from being once-in-a-lifetime bonds between pairs of virgins to being acquired and tossed away like tissues—is a flawed one. Dating as we know it may be a recent invention. But, frankly, so is marriage as we know it. And courtship has morphed and changed and reverted about as many times as I've had to add a contact named "Dan OkCupid" to my phone.

Humans weren't always the marrying kind. Our earliest recorded evidence of matrimony dates back to around 2350 BC in what's now Syria.[2] We don't know exactly why those first folks started getting hitched. One prevalent theory is that, in the days before genetic testing and birth control, such an arrangement was the most reliable way to keep track of paternity—which started to matter to families as they settled down to own land and property. Hunter-gatherers didn't have much use for heirs; farmers and merchants were a different story.

Before then, the question of which baby belonged to whom was probably kind of a moot point. We don't know for sure just how often prehistoric communities did or didn't cluster around nuclear families. Some research has revealed signs of mom-dad-kid units that extended out into related communities;[3] other work has revealed signs of egalitarian groups where sex partners varied and kids were reared indiscriminately by the whole crew.[4] And hey, not to put too fine a point on it, but prehistory lasted a long time, and the earth is a big place, so *por que no los dos?* The point is, we have proof that other ways of life can work.

One solution to the who's-your-daddy conundrum is called partible paternity. This is the notion that more than one man can father a single child. (Before you scoff, know that the scientific world's knowledge of conception is *embarrassingly* recent. We'll get into that more in later chapters. But rest assured that no matter how learned you think your ancestors were, they probably thought human sperm floated on the breeze and pollinated flowers until a couple hundred years ago, at best.) A study of 128 societies in lowland South America (indigenous peoples of the Amazon basin and circum-Caribbean) in 2010 found that some 70 percent of them subscribe to the belief that multiple sex acts with multiple partners are required or preferred in the creation of a child.[5]

The particulars vary. In some cultures, it's thought that babies are literally built out of sperm, 3D-printer-style (this was actually a common belief in loads of places until recently, as we'll learn later). Insofar as sperm is understood as a building material, a woman must acquire lots of it over the course of months or years until she successfully amasses a child. In some societies, while folks believe multiple men can contribute to a conception, they accept that you can *also* make a baby with just one partner. Even these children often grow up considering their mothers' additional lovers to hold second-string paternity roles, perhaps just in case one of them played a part. There is reason to believe that having an extra pop or two can increase a child's likelihood of survival, probably thanks to the potential for additional food and protection—not unlike the logic of the grandmother and gay uncle hypotheses, albeit without an obvious genetic benefit for each individual dad.[6]

So, we know that in some places—and presumably in other times—it's been normal for a child to have multiple fathers. This would, for those in the back, make the OG purpose of "marriage," as it is commonly understood, moot.

In other cultures, moreover, it's normal for a child to have no father at all, at least in a way. The most famous of these are China's Mosuo people, who traditionally practice a form of "walking marriage," which outsiders have long gawked at (and misrepresented).[7] The matriarchal group keeps female lineages living together. This means that male partners generally keep to their own mothers' houses. Contrary to popular belief, most Mosuo women are serial monogamists. They generally know who has fathered their children and encourage some kind of paternal relationship. But breaking up is as simple as ceasing to see one another, no matter how many years or offspring a couple has between them, and the bulk of a dad's typical everyday parenting duties is usually left to male relatives on the mom's side. Many young Mosuo are choosing to assimilate with Han Chinese culture and move to cities, so it's not clear how fiercely future generations will hold on to this unique familial structure.

Then there are marriages that include more than two people. You've almost certainly heard of polygamy. But chances are that what the word calls to mind for you is actually *polygyny*—the practice of men taking multiple wives. It makes sense that this is so common as to border on mainstream (I'm not arguing that polygynous families don't face stigma and even legal punishment in most countries, but once there's an HBO series *and* a reality show about sister wives in the mix, taboos start to look the slightest bit less taboo) because it fits into that old-school, Mesopotamian drive for marriage: a man needs to keep the intended mothers of his children on lock if he wants to be sure *their* offspring are also *his*. The deal, as marriage stood for thousands of years, was a guarantee of legitimate heirs (or at least a much better shot at them) in exchange for a guarantee of financial support. In that paradigm, it's not hard to see why guys with more expansive coffers might want to get multiple irons in the fire, especially if the deal includes a dowry for each bride.

But while polyandry—the practice of women taking multiple husbands—may be less common, it did and does happen.[8] Fraternal polyandry, in which brothers share one wife, makes a particular amount of sense in certain scenarios. Tibetans, for example, traditionally did this to avoid having to divide family plots of land as the generations went on. Several First Nation cultures, including the Shoshoni of central and eastern Nevada and southern Idaho, practiced polyandry as well, perhaps to pool resources and raise the likelihood of a family line continuing.[9] Other cultures around the world have at times accepted the addition of unrelated second husbands to act as "assistants" for the first, sharing in household labor and hunting and gathering.

I'm not big into the "humans were never meant to be monogamous because of evolution" argument. By the same logic, I should have died one of the five hundred times I got strep throat when I was a kid because of evolution. Humans can do things that our ancient ancestors didn't come up with; sometimes natural isn't better. Still, it's worth knowing that our closest relatives in the animal kingdom do things more like the Mosuo, but cranked up to eleven.

Chimps are often stated to be our closest living cousins, and they most certainly are. But most folks hear that and think of the common chimpanzee: a fairly violent critter who speaks to our deep-seated belief that the "primal" undercurrent of our species is fierce and deadly. But there are actually two types of chimp, and in terms of kinship to humans, they're tied.

The bonobo of central Africa is divided from the more famous chimp species by the Congo River and about one to two million years of evolution. Their behavior, also, is radically different. Bonobos tend toward the matriarchal, for starters, with many population groups led by clusters of aged females with long-lasting social bonds. The highest-ranking male bonobo in a group doesn't get to the top by lording his strength over other members; in fact, research suggests violent harassment by males is often met with the formation of female "coalitions" to beat the offender down.[10] And one study found that bonobo males who fostered friendships with females fathered the vast majority of offspring in their communities. Instead of machismo, it is a male bonobo's mother that begets his status. One study even found that mothers watch over their sons while they mate with local females and in some cases will physically drag would-be male competitors out of the way.[11]

This is just to say, I guess, that there are better ways of keeping track of who's responsible for feeding which babies. You can just decide that babies are generally worth taking care of and go from there. Still, the institution has persisted—and led to some very goofy traditions.

HUMAN MATING RITUALS

According to some historians, brides in ancient Sparta shaved their heads and dressed in drag to prepare for their wedding nights.[12] Some experts—including a few men in the 1960s who seem to have been very preoccupied with a need to prove that all the gay sex the ancient Greeks had was completely *not gay*—have suggested that this ceremonial gender-bending was meant to ease husbands into married life. Nearly all male Spartans participated in the *agōgē*, a period of military education that began in early childhood and lasted until they reached full citizenship at the age of thirty. During this time (and after, for as long as they stayed fit enough to fight), Spartan men lived with herds of other boys and men and slept in their barracks. This led, as you might expect, to some amount of acceptance of same-sex activity.

Whether the idea was to make marital intercourse seem more similar to the gay sex Spartan men had become accustomed to or simply to gently introduce them to spending time with women—something they likely hadn't done for more than a few minutes at a time since early childhood—some historians have mused that shaving a woman's head and putting her in a manly robe was intended to make her husband more comfortable in bed. He didn't have to get *too* comfortable, mind: men continued to spend most of their time in the barracks until they retired out of military service or died, and it's said that a marriage could yield several children before spouses ever saw each other in daylight.

But the relationship between homosexual activity among Spartan men and the cross-dressing ritual in women is pure speculation, and it's possible the tradition was rooted in a much more commonplace cultural meme. It may have been a form of apotropaic magic—something intended to confuse evil spirits and keep them from encroaching on the happy day. There are records of women in Argos affixing fake beards to their faces before marriage for similar purposes. While solid sources on this are basically nonexistent, it's often said that bridesmaids originated in ancient times to confuse prowling spirits. It's also possible that having multiple decoys in similar attire served to protect the bride from more practical threats, like spurned former suitors and rival clans, but the "evil spirit" explanation gets more airtime.

Here's a short list of some of the other wedding traditions that, according to top-ranked Google results, likely stemmed from a fear of bad vibes:

- Wearing a veil
- Holding flowers
- Carrying a bride over the threshold
- Smashing a glass
- Ringing church bells
- Painting yourself with henna

But the true origins of most of these rituals remain obscure because the origin of marriage itself is pretty murky—not to mention complex and multifaceted, since it definitely didn't suddenly crop up in one spot and then proliferate in a straightforward fashion. A lot of the things that many of us take for granted as "traditional" for a wedding stem from recent history; white wedding gowns, for instance, became trendy when Queen Victoria wore one in 1840. She likely wore white because it broadcast wealth and power (after all, most folks lacked the facilities to keep fabric looking so clean and bright). Other brides in the sphere of European influence followed suit because Queen Victoria was basically all of the Spice Girls at once. Today, while most of us realize that white isn't a universal wedding color—red is more popular in India and China, to name just one example—it's still considered a "traditional" sign of purity. But until Victoria changed the fashion, blue was the color most associated with feminine beauty and chastity. In fact, it remained the trendy color for dressing newborn baby girls in until around 1950, when it rather inexplicably swapped places with historically manly pink.

What I'm getting at here is that weddings and their origins are kind of arbitrary and complicated. On the one hand, I hope this empowers you to go truly buck wild if and when you ever orchestrate nuptials of your own. Please refrain from appropriating cultures not your own (this is, in general, a tacky move at best), but know that if you take your own ethnic lineage and that of your spouse and spin the wheel enough times, you could probably justify doing your ceremony in drag and a massive black veil as being "traditional" and "respectful of your ancestors" and "absolutely not an embarrassment to the family."

On the other hand, the ways in which we choose to get married are far too myriad and complex for me to really drill into in this book—and weddings themselves don't have much of anything to do with sex. What we do before and after saying "I do" is a different story.

To give a quick summary, it's important to remember that so-called traditional marriage was probably created for pretty boring and gross reasons, like maintaining or amassing wealth or keeping tabs on which babies were yours. Different cultures have had different ways of going about that. What's normal to you might not be normal to someone else, and there's no evidence that the nuclear family has a particularly deep evolutionary origin.

However you feel about matrimony given all this new information, just be glad we've got less freaky ways of preparing to co-parent than some other members of the animal kingdom.

TINGLING SPIDEY SENSES

Like most Australian wildlife, the redback spider, or *Latrodectus hasselti*, is a horror; it's highly venomous and known as the Australian black widow. Luckily, we have antivenom to save you from its bite. But nothing will save you from the terrible feeling you'll get when you learn how it mates.

Sexual cannibalism is when one partner—almost always the female—has a tendency to munch on the other at some point in the process. There's some debate over why this happens, and there's likely more than one explanation. Common theories hold that while a male may miss out on future mating opportunities due to being, well, eaten, his willingness to be, in the words of Lizzo, not a snack but a whole damn meal will make his offspring more likely to succeed. Why did I eat your father? Well, I was eating for two! Makes sense. In some species, aggressive females may simply be more likely to survive and have lots of babies, and that aggression may express itself in some weird ways (like eating the guy trying to fuck you).

Whatever drove the redback spider to cannibalism, the males of the species have a unique flair for the dramatic. Once mounted to deposit sperm, he'll do a sort of somersault so that his underbelly is presented to the much larger female's mouth.[13] It's unclear how many times his intended will say, "Oh no, I couldn't possibly!" before taking a bite, but most dudes who throw themselves onto the proverbial dinner table don't live long after the mating session—if they even survive the whole process.

Why offer yourself up for sexual slaughter? Mates are hard to come by if you're a redback spider, so males risk dying before they can even find someone to reproduce with. That, coupled with the fact that there's *always* a chance your mate will try to take a bite out of you, may have led to an interesting strategy of self-sacrifice. Getting yourself all situated—with sexual organs in just the right spot to make the magic happen—and then offering your lady a veritable buffet of spider booty might make her less likely to disrupt your sperm deposit. If you offer your date snacks before you Netflix and chill, they're infinitely less likely to bail on your make-out session to go rummaging around in your kitchen (or eat you). Ergo, suicidal somersaulting spiders. Evolution is weird.

Some spiders try to get around sexual cannibalism in other ways. Males might bind the female up in silk before trying to mate or present her with a nuptial gift—another bug to eat—in the hope that she won't finish and go looking for more while sex is in progress. Others, like *Argiope aurantia*, are also self-sacrificial, though in a less flamboyant manner than redback spiders. These little guys push sperm into a female with tiny appendages that look like boxing gloves.[14] Once the two mittens are inserted, the male spider just . . . dies. It's a preprogrammed thing. In this scenario, the female may eat her deceased mate (waste not, want not), but his spermy little boxing gloves will remain wedged in her reproductive openings to make sure his swimmers make it— and to make it harder for anyone else to get a shot with his well-fed lady.

THE WORST LIVE-IN BOYFRIENDS
IN THE ANIMAL KINGDOM

Once upon a time, scientists found themselves puzzled by their inability to find male anglerfish. All the specimens they trawled out of the deep sea seemed to be female. Plus, the ladies were riddled with strange parasites they couldn't quite identify. They eventually killed two birds with one stone, so to speak, by realizing the hangers-on were actually the males of the species.

Certain types of anglerfish mate for life in the worst way possible. In the most eldritch display of codependency in the animal kingdom (at least I hope), males bite their chosen mates, ooze digestive enzymes into the wounds, and fuse themselves to their flesh. The female's body provides nutrients to keep her new live-in boyfriend technically alive, but he shrivels and atrophies until he is little more than a nut sack trailing through the sea. This is not hyperbole. He is truly just a living nut sack. Trailing. Through the sea.

Some species of female can carry as many as eight of these lay-abouts at once, using sperm from any of them to fertilize her eggs as she goes about her business. This grotesque courtship likely evolved due to the relative size and emptiness of the deep ocean. If you're an anglerfish, it's so odd to come across another member of your species that it makes more sense to latch on for life than to hope you'll encounter a second mate in the future. Plus, you know, it's either that or keep eating mac and cheese in your bachelor pad.

Anglerfish may give up more than just the potential for promiscuity when they hook up. A 2020 *Science* study found that some species—the ones that go for the most permanent and multipartnered couplings—give up a part of their immune system to do so.[15] The closer an anglerfish species gets to subjecting females to lifelong attachments with eight dudes, the weaker their adaptive immune system gets, at least based on which genes scientists see them deleting.[16] This is no doubt helpful when you are trying to permanently skin-graft yourself to your spouse. A strong adaptive immune system, after all, would reject such a connection faster than Kim Kardashian rejected Kris Humphries.

PORPOISEFUL ACROBATICS

You know a species is real freaky when a scientific paper about its sex life refers to "lateralized and aerial behavior."[17] *Phocoena phocoena*, a dolphin-esque porpoise found along coasts and in rivers across wide swaths of the planet, is one of the smallest marine mammals on earth. It's also into some extremely weird (and difficult to execute) sexual positions.

Researchers observed dozens of mating events from San Francisco's Golden Gate Bridge. Male harbor porpoises, they saw, get lucky by rushing potential partners as the females surface to breathe. They always come in on the left side, according to the

2018 study: an "extreme laterality in sexual approach" that "has not been reported for any cetacean."

They also approach "with force and speed," resulting in "male aerial behaviors" in 69 percent of mating attempts. (Nice.) In other words, they wait for females to leap out of the water and *then* try to hook them with their penises. This often requires that they, too, get airborne.

The harbor porpoise's technique may be particularly shocking, but whales and dolphins in general have fascinating sex lives. Like ducks, many species of dolphin come with intricately designed vaginas that may help them decide which copulators turn into daddies.

Fun fact: Diane Kelly, a senior research fellow at the University of Massachusetts, Amherst, has helped revolutionize our understanding of dolphin sex. And she did so by making 3D silicone molds of their vaginas and *inflating their penises* with beer kegs full of saline.[18] Previously, research had relied on tissue samples from the genitals of dead marine mammals, which were too shriveled and flaccid for researchers to get a great grip on how they acted during sex. Thank you, Diane!

BIRDS OF A FEATHER

Looking for a wingman? Try asking a swallow-tailed manakin. These birds work in groups of two or more males to create little display areas for their courtship songs and dances called leks.[19] One male takes on the role of alpha, perching from the highest vantage point and taking the lead on singing to attract females. His companions will work to build the little lek area and help impress incoming ladies with their moves and tunes—but only the alpha actually gets to mate. The rest of the crew flits out of the way so the couple can copulate.

It's thought that the support team is motivated by the possibility of swooping in if something happens to the alpha. If this is your only motivation for agreeing to cruise the club with your buddy, please just do everyone a favor and stay home.

GENERALLY CRAPPY BEHAVIOR

Here, for no reason other than that I think you should be aware of them, are just a few of the animals that fart, pee, and poo on one another before getting nasty.

- Giraffes: males stick their faces into a female's rear end and smell or taste her urine to make sure she's fertile.
- Hippos: according to some sources, males spray their poop like sprinklers to set the mood.
- Crayfish: female urine is such a crucial aphrodisiac that researchers are able to shut down the whole mating process by keeping them from peeing.
- Porcupines: males essentially super-soak the females they wish to woo with high-speed projectile urine.
- Dung beetles: I mean, duh, but some species will form little balls of poo and then let their lady loves hop on for a ride as they push the feces home.

There's absolutely nothing wrong with being into consensual activities that involve nonpedestrian body fluids (and, uh, solids, hopefully, depending on what you've been eating). But I think we can all agree it's a good thing that these tactics aren't common practice for all humans—singles' night at the bar would be one giant biohazard.

I can't speak to the motivation of every amorously defecating hippo, but it's likely that when nonhuman animals seem determined to cover one another in their stinkiest emissions, pheromones are at play. Pheromones are chemicals that many, but not all, organisms produce that trigger a particular behavior or physiological response in other members of their species. Sex pheromones are those particular chemicals that play a role in getting animals turned on, helping them find one another, and triggering the physical processes necessary for mating to occur. Male cecropia moths, for example, can use the scent of female pheromones to find them from miles away.[20]

While you've probably seen seedy newspaper ads promising you magical sexy powers thanks to bottled human pheromones, that whole concept is bull. Researchers have worked hard to find human hormones that get us horny when we sniff them out on others, but no one has succeeded to date. And it's unlikely they ever will, since the structure that other animals seem to use to pick up these scents—the vomeronasal organ—is vestigial and nonfunctional in humans.[21] Still, research confirms that body smell is a huge part of what makes us find one another attractive.[22] Some studies suggest that we're more likely to think someone smells hot if they're less genetically related to us, giving us an unconscious clue as to how much diversity we'd be adding to our lineage if we made babies.[23]

But there's more to attraction than our ability to procreate (or else we'd all be straight and spend a lot of time sniffing pits). Research that uses computer algorithms to crunch the numbers on relationship outcomes has pretty much confirmed that humans may *think* they know what they want in a partner but that all bets are off when it comes to actually meeting up and hooking up.[24] How we meet, where we meet, what we look like in the moment, and what direction the evening goes in: all of these factors seem much more important than whether two people are good for each other on paper.[25]

Is this freeing? Slightly. Daunting? Extremely. If you're feeling perpetually unlucky in love, I'm sorry to report that no algorithm or secret exists to give you a shortcut to finding romantic bliss. But with attractiveness being such a maddeningly subjective metric, we also know for certain that there is surely someone out there for everyone given the right circumstances.

IT'S A DATE

In the times and places where marriage has been about economic advancement and family politics, courtship has been about proving your worth as one half of a contractual obligation. That's boring. We're not going to talk about that.

Yes, people have no doubt married for love even in times and places where that really wasn't what marriage was for. If nothing else, this book should make you understand that what is "normal" never encapsulates what everyone is actually doing. Likewise, people have likely flirted—within and outside marriage, to be sure—for as long as sex has been a thing we like to do.

But, for most people reading this book, flirting isn't something you do with your chambermaid while your husband looks the other way because he doesn't think lesbians exist. For most people reading this book, flirting is what you do to find a partner in crime—to cuddle, to screw, to live happily ever after with, or just to briefly flirt with, because that can be its own whole thing. And the idea that wooing and swooning is part of the acceptable process of finding a spouse is quite new.

One of the first hints we get of Europeans allowing their kids to flirt comes in a very uncomfortable form: bundling.[26] From the sixteenth through the eighteenth centuries in Wales, Scotland, Ireland, Holland, and the American colonies, it was apparently common enough practice for courting teens (and sometimes strangers) to share beds. This may often have been due to pure logistics: beds were scarce, privacy wasn't really a thing, and if a traveler rolled through town, you might find yourself sharing your space with them for a night. But some records indicate it was sometimes intended as an opportunity for future spouses to get to know one another. One can easily imagine it might also have been a way for parents to test the trustworthiness of suitors their daughters had selected themselves.

In any case, it was a whole to-do: the horny young bedfellows were often tucked or even sewn into bag-like garments, sometimes with a bolster or even a board placed between them, to prevent any funny business. This wasn't foolproof by any means, and many of these young folk likely wiggled free of their bundles to fool around. But in most cases, that would simply speed up the marital proceedings. After all, parents were keeping a tight lid on which men had access to their daughters and when. So those who found their way around the bundling board wouldn't have much plausible deniability if their sweetie fell pregnant.

Love marriage became more of a universal aim for Europeans during the Victorian era. The rise of the middle class and the emergence of new money thanks to industrialization meant some social mores were bound to be shaken up. And the fact that the eponymous queen of the era was known to have selected her husband for love (albeit from a pool of politically acceptable candidates) couldn't have hurt. But proto-dating practices like bundling were viewed with horror.

Marriage may have been, increasingly, about finding love. But, at least for middle- and upper-class folks, it was still arranged under the watchful eye of family members and played out in the presence of chaperones. In the 1870s, in fact, the rising popularity of roller rinks caused some amount of tut-tutting from adults, who worried what young men and women might get up to while recreating in such close proximity.[27] Purity was highly praised in a young woman, and one was generally considered "ruined" if she had sex before marriage.[28]

(This is a good place to mention, by the way, that while many cultures have prized vaginal virginity throughout history, it's highly unlikely that women in these societies were all virgins at marriage.[29] While expectations of bleeding during first-time vaginal intercourse date back to ancient Greece, we know that by the tenth or eleventh century at least, women's health texts advised brides on how to fake such a display. One writer suggested that the bride place a leach on her labia to produce a scab that would be dislodged during sex or else plot to have her wedding night fall during menstruation. This advice would have been particularly crucial given that bleeding and pain during first-time intercourse, while not uncommon, is generally avoidable if the penetrating partner is gentle and the receiving partner is enjoying themself. While some people have hymens that partially or totally cover their vaginal openings, which may tear painfully with pressure from a penis or even a tampon, most people have mucus membranes that simply *surround* the vaginal opening, which may not have to stretch much at all and certainly don't have to rip. It was actually only in the 1500s that physicians started to tout an unbroken hymen as a key sign of virginity; until then, other explanations were given for the expected bleeding, and tests to prove a woman's virtue generally centered on either religious ceremonies or analysis of her urine for various properties.)

Going *out* with someone wasn't really a thing until the 1920s, when young people freshly shaken up by World War I pushed the boundaries of lingering Victorian morality.[30] It's probably not a coincidence that the production, transportation, importation, and sale of alcohol was prohibited in the United States from 1920 to 1933. That drove social spaces underground—no more sitting around at saloons or sipping whiskey in a ritzy gentlemen's club—which gave flirtatious couples a place to hang out outside their family homes. Enclosed automobiles, which were just becoming the norm, provided yet another location away from prying eyes. Newspapers from the decade tell us that adults seethed and fretted over so-called petting parties, where "snuggle puppies" would gather to enjoy monogamous cuddling and kissing en masse. In Atlantic City, New Jersey, policemen patrolling the seaside were reportedly told to throw literal ice water on any teens they saw getting too cozy.[31]

In the 1950s, a new abundance of postwar resources and a new emphasis on education ushered in the era of the American teenager. This was a young adult with loads of free time and plenty of years between the stirrings of their first amorous urges and the expectation of their marriage. The 1960s saw the sexual awakening associated with the contraceptive pill, and while the particulars have changed and your results may vary, dating in the United States has been what it is ever since: sometimes a road to marriage, sometimes a road to a one-night stand, and almost exclusively about the feelings of the individuals themselves. And then, of course, the internet came along.

SEX GOES CYBER

Humans have always tried to use technology to find love.[32] Personal ads have been placed about marital prospects since newspapers first became a thing, and San Francisco and Kansas City played host to the first courtship-centric publication in the United States in the 1870s (men paid a quarter for a listing; women got in for free). Turn-of-the-century personal ads may even have created a safe space for queer people to send coded missives.[33]

Even using data to find your match isn't new. In the 1940s a New Jersey company called Introduction collected info from lonely hearts and offered them matches based on key metrics for twenty-five cents a pop, plus a fifty-two-cent registration fee. Stanford students designed the first computer program for pairing up singles in 1959, and in 1965 Harvard students actually started a company based on that same concept.[34] Several other such businesses popped in and out over the next few decades, some of which mailed out VHS tapes of potential matches to keep their user experience on the bleeding edge of technology. Things went online in the 1990s, and in the early aughts companies began to promise optimal matches based on algorithms.

A recent study found that 40 percent of heterosexual couples in 2017 reported having met online. This number has almost certainly gone up in the time since and itself represents a twofold increase since 2010.[35] The ratio was even higher for same-sex couples, some 65 percent of whom had met virtually.

One important thing to keep in mind, as online dating becomes evermore the norm, is that none of these algorithms has actually been shown to do fuck-all for matching people up.[36] If you find it hard to meet people in the meat space—maybe you're queer in a conservative part of the country, or you have some unusual kinks, or you get intense anxiety when trying to talk to strangers—plugging into an online dating site that caters to your needs can be a huge help in finding what you're looking for. And there are obvious benefits to getting to make sure a person is compatible with you on key metrics before meeting.

I personally think that looking for a partner at, like, a bar (?) sounds like a great way to have a miserably awkward evening or accidentally kiss a bigot. I have literally gone on one postcollege date with someone I met IRL as opposed to connecting with on a dating site, and guess what? Didn't marry him. But no website or algorithm can find people you're more likely to feel a spark for than the average person on the street.

We haven't mastered the laws of attraction. We don't even *understand* them. The heart wants what it wants, and sometimes your clitoris wants something else entirely. There's no right or wrong way to go about it, no matter what the increasingly chaotic New York City subway ads for OkCupid may suggest.

Meanwhile, data on the tail end of the millenial generation only goes to show that the way sex factors into our dating lives is always evolving: as young adults, they had a tendency to have their first sexual encounters later and have fewer partners. Gen Z seems to be continuing this trend, with consent and communication becoming better understood and encouraged over the sort of casual, often inebriated hookups once taken for granted as part and parcel of being sex positive.[37]

A few caveats: It's total nonsense to group entire generations together in terms of behavior, because the concept of a generation is fundamentally made-up and deeply flawed (that's a story for another book, sorry, but I promise people smarter than me have provided ample evidence to prove this). And even if we really could say with certainty that Gen Z, on average, is less into "sex" than previous generations, that wouldn't encompass the lived experience of every member of that generation because averages are not totalities. Plus, as I hope this book has already made clear to you, our idea of what counts as "sex" is incredibly heteronormative and archaic, and many "virginal" college students are probably getting off to user-generated erotica on Wattpad and having heavy make-out sessions on the regular. Not having much or any of a particular kind of sex (or any sex at all, for that matter) is not the same as being puritanical.

But regardless of all that, I do want to say that if young folks are avoiding sex when they don't want to have it, I think that's *amazing*. Sex positivity shouldn't be about making sex feel compulsory. I sincerely hope it's going to get more and more common for people to say no to sex because it's not what they feel like doing at that particular time, or in that particular way, or with that particular person. If sex doesn't interest you until your thirty, or doesn't interest you *ever*, or only interests you after you've been hanging out with your potential first partner for years, then following those instincts is *incredibly* sex positive. It's not about how much we do it; it's about making sure we're happy with the amount and kind of sex we're having (and making sure our partners feel the same). This can change over time, or from day to day, or whatever. As long as you're not hurting anyone, you should never feel bad about what you want your sex life to look like.

There's no right or wrong way to find companionship. Be safe. Be smart. Be kind. Don't poop all over them without asking nicely first. And if all else fails? Remember that you don't need a partner to be fulfilled—or even to have great sex.

Chapter Five

WHAT'S THE DEAL WITH MASTURBATION?

In which graham crackers keep you regular and chaste.

WHAT CONNECTS MEL BROOKS, MUSICAL COMEDY, THE IDEA THAT white men can't jump, and the concept of a chastity belt? All of these were introduced to me by *Robin Hood: Men in Tights*. Watching the film as a curious young girl, I found that last plot device—a bulky set of ironclad undies forcibly worn by Robin's elegant love interest—both disturbing and titillating. The concept of being forced to wear a lockbox on one's box is obviously horrifying from the standpoint of both chafing and personal liberty. But the film used the chastity belt—as worn by Brooks's Maid Marian—to give the character a decidedly unmaidenly air of lasciviousness. Maid Marian isn't just being cock-blocked by her chastity belt to protect her from the whims of lusty men; she spends the whole movie making it clear that she's *desperate* to get the dang thing off and get positively *railed*.

Brooks was onto something: chastity belts wouldn't exist if men weren't intrigued, aroused, and terrified by the concept of fiery female lust. Or, to be more accurate, our *idea* of chastity belts wouldn't exist without the male fantasy of horny lasses in need of protection from their own desires. And that's because, despite how much we've heard about chastity belts, the devices themselves probably didn't exist *at all*.

As anyone who's ever had to take care of a vulva will tell you, there's simply no way medieval missies could have survived being kept in such cages for days at a stretch, let alone the weeks and months they supposedly spent confined while their fathers or husbands went off to war. I wouldn't wear a *bathing suit* for more than six hours without reliable access to Monistat after the fact. In an age before antibiotics and antifungals, a chafing, unbreathing, unremovable undergarment would have caused horrific (if not deadly) wounds and infections.

The idea that medieval maidens were forced to literally gird their loins stems far enough back into antiquity to have an air of truthiness in the modern day. Several engravings and woodcuts from around the sixteenth century depict women in fortified undies, and museums around the world showcase aged specimens made of iron.[1] We even have an illustration of a frightening chastity device from the tail end of the Middle Ages. In the early 1400s, German military engineer Conrad Kyeser penned a tome titled *Bellifortis*, one of the earliest known treatises on military technology. "This is an apron worn by Florentine ladies, made of iron, and hard, to be locked from within," Kyeser wrote, according to a translation by

Albrecht Classen of the University of Arizona. But Kyeser didn't cite his sources, and there's reason to believe the sketch wasn't quite as earnest as some of his military inventions. In his 2007 book *The Medieval Chastity Belt: A Myth-Making Process*, Classen points out that *Bellifortis* includes several "highly fanciful objects," including plans for a device to turn someone invisible.

Classen and other scholars now generally agree that Kyeser was either misinformed or making a gag at the expense of the good people of Florence. And without this questionable reference to chastity belts as real objects, the contemporary citations for their existence drop down to nil. Christian history is, unsurprisingly, rife with references to shielding oneself against sin and calls for young people to don armor against temptation. But you'd have to make several large logical leaps to turn those metaphors into evidence of physical chastity belts.

Instead, Classen and others now posit, those sixteenth-century references to locking up one's lovers and wards were likely examples of our oh-so-common habit of treating our recent predecessors as backward and hilarious. Many of the "medieval" chastity belts formerly exhibited in museums have been reclassified as eighteenth- or nineteenth-century fakes (and in some cases, apparently, sheepishly relabeled as dog collars).

But while these medieval antisex toys may have been mostly mythological, efforts to police a person's interactions with their own nether regions are far too real—and recent. There was nothing metaphorical about pleas for young people to protect their virtue in the 1800s: Victorian-era and early twentieth-century men ran the risk of actually having their genitals squeezed into armored funnels and spiked cages.

But in most cases, it wasn't sex that their guardians and physicians feared. When true chastity belts hit the scene, their aim was to stop masturbation in its tracks.

And the Victorians were right to turn to such drastic measures. I mean, they weren't *right*—there's absolutely no reason to keep someone from masturbating unless they're doing it in front of you and you'd like them to go do it somewhere else instead. But physicians and parents were right to assume that it would take a truly draconian device to put the kibosh on self-gratification. Because many people—like our horny Maid Marian—frequently seek out the pleasures of sex, whether with others or by themselves. And when humans have sex for pleasure (whether alone or together), they're in good company. Let's take a step back to address just how natural it is to want to get off.

OUR HORNIEST COUSINS

Burping the worm. Badgering the witness. Petting the cat. Paddling the pink canoe. Humans aren't unique in their desire to have solo sex. Let's revisit our close cousins, the bonobos.

Bonobos are best known not for the fact that they make nice but for the way they tend to do it: the exchange of sexual favors in a bonobo colony rivals the most free-lovingest 1970s hippie commune and may just put all human attempts at sexual egalitarianism to shame.[2] Bonobos start playing with sex as young as age one—almost a decade and a half before they reach adulthood—and do so with a wide enough variety of positions to refute the notion that they might just be practicing for reproduction.[3] They've even been shown to exhibit this behavior in captivity, with no adults around to show them how to do funny business.

Mature female bonobos engage in sexual activity even when they're not fertile. They frequently rub their genitals against those of other females for what seems to be mutual pleasure. Males, not to be outdone, are known to indulge in "penis fencing": hanging from the trees and swinging their genitals into one another for stimulation. Males and females alike have oral sex in various pairings. They kiss with tongue. They masturbate. They use a position somewhere between missionary and cowgirl about a third of the time they have penis-in-vagina sex, according to one study.[4] This setup is particularly neat, since very few primates are known to do it with partners facing one another.

Here are a few other examples of pleasure-seeking individuals in the animal kingdom.

LEAPING LIZARDS

Masturbation isn't always about trying to get your rocks off, as it were. Sometimes it's about getting off on a rock so you're more likely to become a dad. Let me explain.

For male marine iguanas in the Galápagos Islands, spreading one's seed can be an extremely difficult enterprise. Each female will only hook up with one guy per mating season—and she'll only give him one shot. That level of competition would be tough enough, but the process is further complicated by the fact that males need around three minutes of thrusting to finish. During that time, it's common for rivals to attempt to shove the penetrator out of place. If you're a big dude, that's not such a big problem—you'll generally be able to stand your ground until you orgasm. But if you're on the puny side, you face a high risk of getting punted.

Researchers have observed smaller males using a brilliant tactic to get around this issue: masturbation.[5]

When boy meets girl, instead of hopping right over to do the deed, he'll have a solo session first—taking a copulatory stance against a rock and ejaculating. Then he'll store that semen in the little pouch that houses his penis. It can stay viable there for hours.

Now, when he thrusts into a female, she'll get an immediate dose of spunk. Even if he fails to finish inside her, there's a solid chance that his sperm—and not the sperm of whatever interloper knocks him aside—will fertilize her eggs.

SOMETIMES MONKEYS HAVE SEX WITH DEER

Besides the genital-rubbing bonobos, loads of other primates are into wanking too. There's one 2017 study on this in particular that I just can't get out of my head. In it, researchers described a hitherto unknown instance of interspecies intercourse: adolescent female Japanese macaques mounting sika deer for the sake of sexual pleasure.[6]

As best as the researchers could guess, this was a new adolescent trend. Japanese macaques are known to occasionally hitch rides on sika deer, who are apparently very chill about this as it's an opportunity to get groomed in areas they can't reach. Meanwhile, female monkeys are also known to mount their primate peers during adolescent years as a means of sex play. It seems like some brave little macaque put two and two together and started relying on deer for her masturbatory needs, and the practice caught on among her fellow teens.

Male rhesus macaques are known to masturbate when they're having trouble competing for female attention. In fact, some research suggests that ejaculating is not the name of the game. Instead, like lizards, they masturbate to get themselves *ready* to ejaculate, in case they get a brief window of opportunity with a mate who would otherwise be out of their league.[7] Yakushima macaque males do so with such frequency that at least one researcher realized they could rely on self-supplied semen for study samples instead of coaxing it out in various distasteful ways.[8]

BEEN THERE, DONE THAT (SOLO EDITION)

If the scandalizing self-love antics of monkeys and apes are any indication, masturbation goes way back in our evolutionary history. And this means humans of all sexes and genders have likely been doing the five-finger shuffle since the very beginning.[9]

Indeed, several cultures feature creation myths with masturbatory aspects. In ancient Mesopotamia, the Sumerian god Enki's power was closely tied to his semen. According to some translations of religious texts about his goings-on, his self-emissions helped pollinate the world with life and formed the Tigris and Euphrates Rivers. Ancient Egyptians sang hymns to the god Atum recounting the solo act of creation—where he either masturbated into his mouth by way of his hand or managed a more direct deposit using godlike flexibility—that allowed him to give birth to other gods and, by extension, all living creatures.

Even when they didn't credit the creation of the whole dang world to self-love, ancient cultures seem to have generally taken a positive to neutral stance on it. Ancient Greek pottery frequently depicts satyrs engaging in group masturbation. Evidence of Greeks' real-world practices is scarce, but references in the works of Aristophanes suggest, at worst, a kind of good-natured disdain for men who took matters into their own hands.

The practice might have been the butt of jokes and was often associated with the lower classes or men with little self-control. Even so, masturbation was likely seen as a perfectly reasonable alternative to sex with another person. The issue in ancient Greece wasn't that you masturbated; it was the implication that you masturbated because you couldn't get anyone to lend you a hand. Even one mythological origin story for the practice hints at this tension: the Greek god Hermes was said to have taught his son Pan how to masturbate after a nymph spurned his advances, at least according to a somewhat joking tale recounted by the philosopher Diogenes. Pan then went off and taught all his favorite shepherds how to do it, presumably while shouting "no homo" at regular intervals.

Speaking of Diogenes: when he allegedly took to masturbating in public, claiming that no natural human act should be considered shameful, his neighbors disagreed. But I think we can all recognize that this should be viewed as a sign not that masturbation was taboo but that Diogenes was being a total jerk about jerking off. Ancient Roman references follow in the same vein, with masturbation treated as a normal but kind of loserish way of seeking sexual release.

The Laws of Manu—a text from around the second or first century BC that set moral and social standards of behavior for most practitioners of Hinduism—bemoaned the loss of potentially procreative semen and advised men on how to atone should they self-stimulate to the point of orgasm.[10] There are also, however, secular texts from the same era and even religious tomes for particular Hindu sects that encourage masturbation. This suggests that only men who took vows of chastity or were particularly conservative in their religious beliefs actually adhered.

Taoist tomes from ancient China advise men not to ejaculate thoughtlessly lest they lose too much yang, the essence of their masculine energy. But there wasn't an outright stance against masturbation: men were simply advised to not ejaculate during most wank sessions. Some Taoists maintained it was best never to ejaculate at all, even during sex, though others suggested it was fair game after one's female partner had experienced multiple orgasms; the excess of her yin would help restore the male's lost yang.

Speaking of ladies, you may be wondering where the heck they are. History just hasn't kept great notes on non-penis-having masturbators, presumably because history is written by the victors and most of those victors had dicks. One historical overview from 2002 notes that in ancient China, to name just one example, the historical record suggests scholars simply "tolerated or ignored" female masturbation, whereas the male act inspired loads of discourse as to pros, cons, and methods.

We know, of course, that all humans have always masturbated (I mean, not *all* humans, it's fine if you don't wanna, but like, statistically speaking, all humans). Just circle back to our chapter about the evolutionary history of sex and read up on what bonobos get up to if you don't believe me. And while most historical texts about sex focus on male desires and acts, there are indeed little sprinkles of references to women masturbating throughout history.

Ancient Greek pottery sometimes depicted female figures enthusiastically using dildos, for instance. In the early ADs, the

Greek physician Galen suggested masturbation as a way to trick a baby-hungry womb into satisfaction so it wouldn't start to sicken virginal or widowed patients.

In the late seventeenth century, a writer claiming to be a female nun in Japan provided in-depth instructions for when and how to masturbate while cloistered.[11] While it fits the general paradigm of that time and place in assuming that only women deprived of men would be interested in self-pleasure, the techniques it describes do seem earnest, as opposed to just being conjecture fit for the male gaze. The writer advises her audience to practice masturbation in the privacy of the toilet at first, to make sure they're able to control themselves well enough to avoid detection when masturbating alone at night or during a lull while working. She also notes that, whenever possible, women should collect and ingest their ejaculatory fluid so as not to waste their energy.

In early modern England—around the 1600s—poems, plays, and porn intended to titillate used female masturbation as a common subject, and texts written for midwives seem to confirm that this actually was a widely known, if perhaps not widely accepted, practice.[12] Nicholas Culpeper's *Directory for Midwives* notes that women may "itcheth" in a lustful manner and stimulate themselves with their fingers. He goes so far as to advise new husbands not to fret if their brides don't bleed on their wedding night, as it's likely they simply surrendered to the temptations of masturbation.

But the references to female pleasure in most of history are few and far between, at least relative to the male equivalent. Even in anti-masturbation literature (more on that later), women said to overmasturbate were often described as mannish in their proclivities—enlarged clitorises were often blamed for a young lady's desire for stimulation.[13]

A BUZZWORTHY TOPIC

There's at least one widely shared story about a female historical masturbator. According to roughly three bajillion people on the internet, Cleopatra used a vibrator made of bees. More specifically, she's said to have ordered enslaved people to fill jars or hollow gourds with buzzing insects and then seal them shut. When she was feeling randy, the story goes, she would just shake the vessel to anger the critters within. *Et voilà!* A vibrating vase full of bees.

Before we get into the sourcing on this supposed origin story for the vibrator, I'd like you to stop and think about it. *Bees.* Just, like, *dried pumpkins full of bees.* The impracticality of it alone boggles the mind. Did enslaved workers go out and catch fresh bees every time Cleopatra wanted to masturbate, or did they keep some kind of horny apiary going behind the palace? Also, how many bees are we talking about here? I have never held a gourd full of angry bees, but I can't imagine they make a very powerful rumble unless they're veritably packed to the gills. And that turns this sexual gourd shaking into a risky proposition for the masturbator, does it not? All it would take for the queen of the Nile to end up covered in bees is one clumsily sealed vibrator (or one particularly enthusiastic shake).

But it doesn't take much digging to show that the origin of this historical tale isn't historical at all, unless you're calling me ancient. Like your fair author, my dear reader, this buzzy story first surfaced in the year of our Lord 1992. No one has yet found a source for it before its inclusion in the *Encyclopedia of Unusual Sex Practices* by Brenda Love. It's time for this implausible vibrator origin story to buzz right off.

Even if you're unfamiliar with Cleopatra's dubious connection to the sex-toy industry, you've almost certainly heard another tale about how vibrators rose to popularity. It goes like this: During the Victorian era, doctors treated "hysteria"—a catch-all term that turned "women having feelings that were inconvenient to the men around them" into a medical condition—with a procedure called "pelvic massage." The goal of this treatment was to induce a "hysterical paroxysm," which we now recognize as an orgasm, to get those hysterical ladies to loosen up a little.

But oh, how exhausted those male physicians were by the increasingly frequent demands of their female patients! Entire clinics found themselves devoted to full-time fingering. Hand cramps abounded. Along came Joseph Mortimer Granville, inventor of the vibrator as we know it, to automate the process and increase orgasmic efficiency across the health-care sector. This story was the subject of a play, which was nominated for three Tony Awards, as well as a movie starring Hugh Dancy. But have I got news for Hugh: it's all made up.

Mortimer Granville did indeed invent a handheld, electric massager in the 1880s. But it wasn't for producing orgasms. In fact, most physicians thought that vibrators were best kept far away from lady bits. Granville himself considered female stimulation to be a misuse of his invention, which was intended to treat pain and nerve problems in male patients. In an 1883 book on the subject, he vehemently denied ever using his device on a single clit.[14] "I have avoided, and shall continue to avoid the treatment of women by percussion," he wrote, "simply because I do not wish to be hoodwinked, and help to mislead others, by the vagaries of the hysterical state."

The idea that Granville sought to coax orgasms out of emotionally distraught women seems to stem from the 1999 book *Technology of Orgasm*. Here, author Rachel Maines cites numerous sources to support claims that pelvic massage was all the rage and Granville's vibrator was used to improve that procedure. But in a 2018 study published in the *Journal of Positive Sexuality*, Georgia Institute of Technology's Hallie Lieberman and Eric Schatzberg systematically unraveled Maines's entire argument.[15] When Lieberman attempted to confirm Maines's claims (something she tried, by pure coincidence, as part of an assignment in graduate school intended to help new historians learn how the research process works), she realized that none of the sources cited in the 1999 book actually contained information about vibrators being used to induce female orgasm in hysteria patients. In response to the criticisms laid out in Lieberman's paper, Maines told the *Atlantic* that she'd really meant for the whole Victorian vibrator narrative to be an intriguing hypothesis based on inference and conjecture—not a confirmed fact.[16] But that clearly didn't stop the world at large from latching on.

Lieberman argued in a 2020 *New York Times* op-ed that this false narrative wasn't just bad for spreading mistruth like wildfire:

If you swap the genders you can recognize how much the widespread acceptance of this story is based on gender bias. Imagine arguing that at the turn of the 20th century, female nurses were giving hand jobs to male patients to treat them for psychological problems; that men didn't realize anything sexual was going on; that because female nurses' wrists got tired from all the hand jobs, they invented a device called a penis pump to help speed up the process. Then imagine claiming nobody thought any of this was sexual, because it was a century ago.[17]

Sexism aside, it's not surprising that the false narrative caught on. There's something wonderfully salacious about the idea of doctors making their patients orgasm so hard that they considered it a cure for anxiety and depression. As Lieberman often points out, it also fits neatly into our pervasive belief that sexuality has evolved linearly—that our ancestors were prudes and that we've achieved the peak of free-loving enlightenment and so have earned the right to sneer at their ridiculous antics.

As this book has hopefully already made clear to you, this couldn't be further from the truth. Our species' history is a loop-de-loop of progression, regression, suppression, and exhibition. Modern-day humans should not assume their sexual norms are the peak of *anything*.

The real plot arc of the vibrator as a sex toy may be less salacious, but it also makes more sense. Yes, male physicians understood that physical stimulation could send women into throes of ecstasy. No, that was not something they wanted women to seek out. Yes, the inventor of the electric vibrator saw its potential as a sexual stimulator. No, he did not want women to use it when they masturbated.

Instead, vibrators rode the same wave of cure-all quackery as radium suppositories (don't worry, more on those later) and other sorts of snake oils. Throughout the beginning of the twentieth century, the devices were marketed for everything from soothing colicky babies to getting rid of hemorrhoids. It's not clear when people caught wise to their sexual potential; given humanity's ingenuity when it comes to finding newer and better ways to get off, one can assume that *someone* was using a vibrator to masturbate from roughly day two of the object's existence. According to Lieberman's 2017 book *Buzz: A Stimulating History of the Sex Toy*, at least one 1903 advertisement referred to vibrating belts as "Sexual Vibrators," and other early twentieth-century ads arguably hinted that the devices might have a lurid off-label use. But it wasn't until the 1970s that their utility as sex toys became part of mainstream discourse (and marketing).

But at the time of the vibrator's Victorian debut, physicians weren't much concerned with helping folks get off. In fact, they were rather preoccupied with getting people to *stop*.

DON'T KEEP YOUR HANDS TO YOURSELF

Unless you're one of a very privileged few (read: raised by hippies or woke millenials), you probably grew up either not hearing a thing about masturbation or hearing that it was bad. And in reading about the relatively permissive attitudes on self-pleasure found in many ancient cultures, you might reasonably start to wonder what the heck happened to our laissez-faire attitude.

One aspect of this tale is religious in nature. As you are probably aware, many spiritual institutions (though certainly not all) take a negative stance on wanking. While the Bible itself doesn't say anything specific about self-love, the Genesis story of Onan is often slotted in as an anti-masturbation anecdote. Onan, Judah's second son, is told to marry his late brother's widow and give her a son. He doesn't actually want to have children with her, as a male heir would take the place of his late brother and jump his spot in the line to inherit a bunch of stinky sheep or whatever. So while he does consummate their marriage, he spills his seed upon the ground.

Oh no! I think he gets smote for that. I'm not going to look it up. It's not important. The point is that Old Testament God says not to pull out. Now, views about masturbation obviously vary between various Christian sects and even their individual members. Still, a quick Google search yielded more than one Catholic think piece about the dangers of "onanism," by which they mean masturbation.

But anti-masturbation sentiment in the Western world has a more recent—and decidedly secular—point of origin. Throughout the eighteenth and nineteenth centuries, physicians published numerous pamphlets and papers on the physical dangers of onanism. The big kickoff is thought to have been the circulation of *Onania: or, the Heinous Sin of Self-Pollution* throughout London and Boston in the early 1700s. It influenced the work of Swedish doctor Samuel-Auguste Tissot, who in 1758 claimed semen was an "essential oil," the depletion of which would lead to all sorts of deleterious health effects. By 1838, French psychiatrist Jean-Étienne Dominique Esquirol confidently printed that masturbation was "recognized in all countries as a cause of insanity."

Here is just a small selection of the conditions on which these gentlemen and their colleagues blamed excessive masturbation:

- Loss of appetite
- Increased appetite
- Paralysis
- Impotence
- Loss of libido
- Weakness
- Vision and hearing loss
- Coughing
- Back pain
- Cognitive decline
- Rage
- Fever
- Insanity
- Organ failure
- Memory problems
- Gout
- Rheumatism
- Headaches
- Blood in the urine
- Neuralgia
- Liver and kidney disease
- Urinary problems
- Uterine cancer
- Epilepsy
- Suicide
- Looking like this guy:

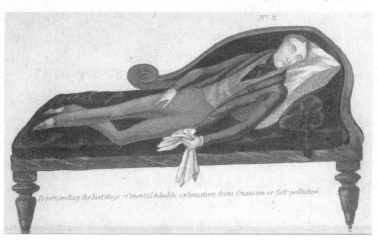

Representing the last stage of mental & bodily exhaustion from Onanism or Self-pollution

The notion that excessive self-stimulation would turn young men into withered Benedict Cumberbatches naturally drove innovation in the anti-masturbation market. Some entrepreneurs took a physical approach, creating all-too-real devices not unlike the fake chastity belts of antiquity. There were also spiked and spring-loaded rings designed to prevent nocturnal emissions by waking a pained wearer who became erect, as well as electrified cages and systems that pumped cold water into the user's pants to put out any flames of desire.

It should be noted that many of the most bizarre objects in this category may never have been created. The US Patent and Trademark Office has records of dozens of anti-masturbation devices from the mid-1800s to early 1900s, and data on which were produced or how popular they became is scarce. While we know that physicians were quite widely peddling anti-masturbation rhetoric to their patients, it's likely that most people used simpler methods, like tying their children's hands and feet to bedposts or keeping them from getting constipated.

MASTURBATION: PART OF A BALANCED BREAKFAST?

We simply cannot discuss masturbation without talking about corn flakes and graham crackers. Neither food was explicitly invented to stop people from jerking off. But both products were born out of health and spiritual movements that placed masturbation high on America's public enemy list.

Graham crackers came to us courtesy of one Sylvester Graham, a nineteenth-century Presbyterian minister credited with popularizing vegetarianism in America.[18] He wasn't in it to save the animals: Graham believed that humans were truly meant to eat only a plain, plant-based diet packed with coarse-ground flour.[19] Graham's vision for America featured women toiling diligently at home to mill and bake their own wheat products, while men worked hard and avoided vices like alcohol, soft beds, and warm baths. Anything too enjoyable was a slippery slope into wanton carnal desire, and flavorful food was no exception.

Graham's teachings dovetailed with other reform movements of the era and picked up enough speed that so-called Graham crackers were being mass-produced by the turn of the twentieth century. The plain, wholesome biscuits offered followers a way to follow the letter of the law without spending the whole damn day milling grain. It goes without saying that s'mores and Teddy Grahams would have given him a heart attack.

John Harvey Kellogg was a contemporary of Graham's with similar goals and strategies.[20] His family were members of the newly minted Seventh-day Adventist Church. Its founders, Ellen and James White, were already preaching the virtues of spice-free living to prevent masturbation by the time they sent Kellogg to medical school to be the religion's chief physician. But Kellogg grew to really, really, really, REALLY focus on abolishing one particular source of sexual stimulation: constipation.

Believe it or not, Kellogg suspected that the pressure of a full, irregular bowel could drive an undisciplined person to sexual distraction. We can reasonably infer that Kellogg, when constipated, was himself aroused by the resulting poo pressure on his prostate, often referred to as the "male G-spot," which is responsible for much of the pleasure brought about by anal sex. At his Sanitarium in Battle Creek, Michigan, patients were treated with brisk exercise, several types of enemas (using high-powered water or even yogurt), and vibrating chairs designed to shake the poop right out of them.[21] They also, naturally, were fed plenty of fiber.

Kellogg and his brother Will worked for years to create a bland, easy-to-digest, ready-to-eat food to help free the masses from the temptations of slow bowels, and they landed on corn flakes—sort of.[22] Will actually had to break ties with his brother in order to coat the cereal with sugar, which he rightfully felt was key to making it marketable. A cereal that makes you poop *so good* that you'll never be horny again may have been the ultimate dream of the Seventh-day Adventists. But it wasn't the sort of selling point Americans were ready to see on a billboard.

The good news about masturbation is that Graham, Kellogg, et al. had no idea what they were talking about. You're only actually doing it "too much" if it starts to hurt you.

If your skin has grown raw and chafed, you should see a doctor for reasons that I hope are obvious, and it's possible that something like a yeast infection is to blame. But if you report multiple jerk-off sessions a day and no other cause can be found, your physician may tell you it's time to cool it on the self-stimulation. Other potential issues include a loss of sensitivity (which is luckily easy enough to fix by taking a break or changing up your technique) or an inability to focus on other aspects of your life.

It's a lot like video game addiction: despite some moral panic around the idea that playing first-person shooters might turn young folks into murderous monsters, many studies have concluded that frequent gaming is unlikely to bring out dangerous behavior in and of itself. And some scientists argue that nothing about playing video games inherently puts you at risk of becoming "addicted" to them. There are exceptions to every rule, and there are no doubt some truly vile video games out there that you'd have to be a little unhinged to want to play (and ones like

Candy Crush that are actually designed to get your brain hooked on leveling up). But gaming itself is probably a neutral activity, from a public health standpoint. It may even be a great way for you to de-stress, explore new ideas, and interact with friends (just like masturbation).

According to the World Health Organization, one *can* become addicted to gaming; the group added "gaming disorder" to its International Classification of Diseases in 2019. But many researchers argue that there's no hourly allotment at which point a gamer is automatically exhibiting disordered behavior. Unlike with, say, alcohol or cocaine, there is no known, inherent harm in regular—or even frequent—use. If your job involved playing video games (or masturbating) for eight hours a day, that wouldn't be a reason for concern (though chafing would probably be an issue in either instance).

Problematic masturbation is much the same: wanking is only bad if it does bad things to your life, like keeping you from tending to crucial responsibilities or negatively affecting your relationships. Playing *Call of Duty* for just an hour a day might be a sign of trouble if you do it even though your infant is screaming to be fed or you have a broken finger that makes every shot agonizing. Playing it for twelve hours straight every Saturday because you've got nowhere to be and that's your primary way of hanging out with your long-distance bestie might actually be one of your healthier habits.

In summary: Are your masturbation practices causing you physical anguish? See a doctor. Does it seem like they might be affecting your work, education, or personal relationships, or do you feel compelled to masturbate in a way that causes harm or distress to yourself or others? See a mental health practitioner. If not, keep calm and wank on. You come from a long line of masturbators, I promise.

And hey, there's at least one major benefit to taking matters into your own hands: you're infinitely less likely to catch something.

Chapter Six

WHY ARE WE SO SCARED OF STIS?

··

In which several venereal diseases are actually found to be quite fucking badass.

··

WE CAN'T TALK ABOUT THE EVOLUTIONARY HISTORY OF SEXUALLY transmitted infections (STIs)* without talking about other places they crop up in the animal kingdom. And we can't talk about animal STIs without talking about koala chlamydia.

I can talk *so much* about koala chlamydia, you have no idea.

---

* Quick vocabulary lesson, y'all: What's up with the shift from STD to STI? STD stands for sexually transmitted disease. STI stands for sexually transmitted infection. The difference might seem trivial, but it's important. An infection is caused by a microorganism that has infiltrated a foreign system. A disease is a set of structural or physiological malfunctions that can occur in a body. Infections sometimes cause diseases, but they don't always do so. As you'll learn in this chapter, many infections—including sexually transmitted ones—are asymptomatic and prefer to lurk incognito in the human body.

Koalas are pretty much universally considered to be adorable. Heck, the public restrooms of America are full of Koala Kare–branded diaper-changing tables. So there's some amount of shock value in revealing that these cuddly critters are, generally speaking, riddled with chlamydia. But while you may feel the urge to guffaw, this affliction is no laughing matter: chlamydia is endemic and potentially lethal in the species, and the antibiotics used to treat them are knocking out crucial bacteria in their guts. Koalas are far from alone in carrying a version of this microbe; it's found in various birds, fish, mammals, and even amoebas. Chlamydia exists on every continent and in the ocean. But koalas have it particularly rough.

A human might suffer from the runs while enduring antibiotic treatment, but a koala—which relies on a delicate balance of gut flora to digest an otherwise toxic diet of eucalyptus leaves—can easily starve as a result of the chlamydia cure.[1]

Unfortunately for koalas, chlamydia is serious business Down Under. With only tiny pockets of the critters left unaffected by the infection, the rest are facing sexually transmitted extinction due to side effects such as blindness and infertility. Climate change has their water sources dwindling and their habitats shrinking with every passing wildfire, and chlamydia—or perhaps the gut-wrenching antibiotics that cure it—could prove to be the kiss of death for this charismatic species.[2]

Humans may not be stumbling around crusty-eyed and helpless as a result of our chlamydia infections. But still, in April 2021, the United States reported record-breaking rates of the disease for *six years in a row*, from 2013 to 2019. Crikey!

And while popping antibiotics isn't as immediately detrimental to our health and well-being as it is for our furry friends, that doesn't mean we can rely on them. The World Health Organization notes that while gonorrhea is the STI most likely to evade antibiotics, chlamydia shows early signs of resistance too. But we have a secret weapon that koalas don't: condoms. If you won't suit up for yourself and your partner, do it for the li'l guys in Australia who wish they could.

Long before humans knew that contact with microbes could cause disease (this, my friends, is called germ theory, and we have a man named Louis Pasteur and some dirty, kinky flasks to thank for it), we knew that having sex could make us sick. Or at least . . . we sort of did. Sometimes. The history of medicine is riddled with descriptions of strange leprosies that focused on the genitals, cruel skin afflictions most assuredly caused by poor hygiene, and oozy rashes sent down from heaven or up from hell, depending on the religious dogma of the time.

But STIs are older than even our historical records. Before we humans were writing down what we thought about STIs, we were passing them on as we arrived in new lands, leaving traces of these encounters in our genes and in the world. In fact, we now know—as koala chlamydia makes clear—that STIs are older than humankind itself. Or as Joel Wertheim, associate professor in the Division of Infectious Diseases of the Department of Medicine at the University of California, San Diego, puts it, "Herpes viruses have been with us since we were fish."

Wertheim got into the historical STI game in a backward fashion—he actually studies modern-day transmission networks to help find ways to stop the spread of infections like the human immunodeficiency virus (HIV) and works closely with the CDC and the City of New York to put his models into action for public health outreach. But in delving into the rapid evolution of RNA viruses such as these, he became equally fascinated with slower-adapting, easier-to-track conditions like the human papillomavirus (HPV) and herpes simplex virus (HSV), the sorts of viruses that have been moving in tandem with our ancestral lineage since the life aquatic. Wertheim waxes poetic on the pervasiveness and tenaciousness of various viral strains, ever amazed by their ability to "exploit a necessary part of human survival" by hijacking the act of reproduction.

We can't really know how our primordial ancestors (and close cousins like Neanderthals and other *Homo*s) reacted to the symptoms of STIs. But here's the thing: we *can* say that they absolutely had them. In fact, a growing body of evidence suggests that humans and other hominid species swapped infections back and forth while interbreeding. Our DNA shows signs of frequent human and Neanderthal romances, yes. But it also shows something else. In interbreeding, early humans transmitted HSV to Neanderthals; not to be outdone, Neanderthals transmitted HPV to us.

But before bitterly blaming these early encounters for today's afflictions, listen to researchers like Simon Underdown, a chipper reader in biological anthropology at Oxford Brookes University. According to him, we're awfully lucky our ancient relatives passed these pathogens around. Thanks to them, Underdown explains, we know what early humans got up to.

Attempts to sequence ancient DNA have come a long way. When Underdown and I chatted in 2018, he cited a limit of around half a million years before fragile double helixes were just too degraded to read. Recent research has managed to peek at the genetic code of mammoths from just over a million years of age. But that still isn't enough time to understand how our species evolved.

Underdown and his colleagues can track the evolution of a virus to see when different strains came into contact with different species of human ancestor or when they moved around different parts of the globe. That provides a sort of shadow record of their comings and goings. And it doesn't just go further back in time than DNA evidence; the viral work also confirms the sexual encounters inferred based on genes seemingly swapped between species.[3]

"The nice thing about our work is that it's very complementary to the recent findings using ancient DNA, which has totally turned our view of early hominin interaction on its head," Underdown says. "We now see evidence of six or seven distinct species intermingling. Our work can look at a broader time window, and can show definite contact even without evidence in the fossil record."

Researchers hope to eventually do this kind of sleuthing (a who-done-who, if you will) with all sorts of viruses. But HSV leaves a remarkably helpful trail of clues due to a few of its unique qualities. For starters, HSV is largely asymptomatic in many of the primates that pass it around, and regardless of pathology, it's rarely fatal.[4] That makes it possible for herpes to proliferate among a population that's otherwise healthy enough to keep living, breathing, fighting, and making out—key to an STI's survival.

HSV also has an ingenious method of infiltration to boost its gregariousness: when a carrier sheds skin cells that contain the virus (a process called "viral shedding") and those particles come into contact with an opening in another person's skin—either in the mucus membrane or by way of an injury, even a microscopic one—HSV attacks potential host cells with astounding voracity. But once it successfully infects its new habitat, HSV travels along nerve cells to live dormant in their roots. The signature tingle of a rising herpes sore that marks the awakening of the microbe can happen infrequently or not at all. Cooties are much more likely to spread successfully if they don't make much trouble for the bodies porting them around, and herpes takes full advantage of that fact.

But wait, there's more: While all other primates, as far as we know, carry just a single strain of HSV, humans have two distinct flavors of HSV. And one is suspiciously similar to the variant found in chimpanzees. That's the mystery Underdown and his colleagues set out to crack.[5]

As is most common in strains found in other primates, HSV-1 prefers to live in mouths; this virus is the source of most human cold sores. But unlike other apes, humans also have a second endemic strain, HSV-2, which notoriously seeks out lodgings further south.

Our history with HSV-1 is simple. Research suggests it's been in the human family for as long as seven million years. That's *before* we diverged from our common ancestor with chimps. Most vertebrates in the world have a strain of herpes that evolved along with them; this is ours. HSV-2, on the other hand, makes us stand out—and tracing it back in our timeline makes clear that the virus is a relatively new addition.

Underdown and his colleagues started with what others had determined by examining the family tree of modern viral strains: genital herpes (HSV-2) jumped from chimp ancestors to the human lineage somewhere between one and three million years ago or so (remember, that's millions of years *after* HSV-1 first got on our nerves). Our ancestors scuttled around hot tropical areas not well suited to DNA preservation during that window. So, for Underdown, finding a great-great-great-great-great (etc.) grandpa with genital herpes was not a likely culmination of this caper, even with recent (and presumed future) advances in sequencing technology.

Instead, the researchers used paleontological climate data to pinpoint where chimp ancestors would have been prowling during the proper period; they then overlapped fossil and climate data for our own forebears, until they found the intersection of protohumans and protochimps. The bipedal hominin *Paranthropus boisei* was in the right place at the right time (or the wrong place at the wrong time, depending on how you feel about herpes) to pick up cooties from chimpanzees by tussling with them or eating their flesh. While you can't trace a straight line from this herpetic hominin to humankind as we know it—*P. boisei* isn't an ancestor of ours but rather one of the many flavors of great ape that failed to produce a lasting lineage—this close genetic cousin would have had many opportunities to interact with and be eaten by our close evolutionary grandpapa *Homo erectus*.

Homo erectus would have already had some amount of immunity to all strains of HSV due to its species' rampant level of infection with so-called oral herpes, or HSV-1. Against such immunities, HSV-2—the new arrival from chimpanzees, by way of *P. boisei*—could only gain the foothold it did by adapting to slightly different mucosal digs. So it sought out safe harbor in our ancestors' genitals.

But, then as now, STIs are always on the move. In fact, HSV-1 is undergoing something of a comeback in order to survive our modern sexual proclivities. Today, more and more people are engaging in unprotected oral sex, when they've had little or no experience with penetrative intercourse (good on ya, teens, but please use condoms and dental dams). Meanwhile, folks have gotten wise to the idea that cold sores are contagious and have generally gotten more conscientious about not smooching babies during oral herpes outbreaks. This may mean that more folks are getting their first-ever encounter with herpes during oral sex. As a result, "oral" HSV-1 is increasingly finding opportunities to jump onto genitals that have never heard tell of HSV-2 and have the lack of protective antibodies to prove it.[6] That means "cold sores" are increasingly causing outbreaks of "bad herpes," and we're left dealing with the semantic fallout.

These ingenious viral infections have a remarkable capacity to give us a glimpse into the past. But studying them has also made Underdown think about how they'll shape our future: namely, how people might stigmatize the diseases less if they understood how intrinsic they are to the human experience. "Infections like this are something that affects all of us," he says, "and have been with us for such a long time."

We might wonder what it was like to get an STI before the dawn of society as we know it. "I do imagine that stigma around symptoms is nothing new," argues Wertheim, "because of course the less you know about something the scarier it is." But it's unlikely a human nomad's reaction to a random genital wart or two would be anything more dramatic than "huh." Of course, HPV can lead to various cancers: that's why all those commercials are out there to guilt you into vaccinating your kids (do it!). So it's not outside the realm of possibility that one of our ancient STI patients would have ultimately succumbed to complications from an asymptomatic infection.

Still, this was a world where getting gored by a mammoth or dying of sepsis after scratching your butt too fiercely was perhaps even more probable. So I think it's safe to say that the ill effects of HPV were relatively minimal in early human society.

If early humans and Neanderthals failed to shame one another for their sexy microbial strains, we've done a great job of shaming them retroactively. *Smithsonian Magazine*, for example, reported on the study linking HPV to Neanderthal sex by saying the hominids gave us "good genes and nasty diseases."[7] Similarly, a 2016 study showed we likely passed on HSV during our Neanderthal rendezvous; researchers suggested this exchange might have made one *small contribution* of many to the hominids' demise. In response, many media outlets framed the research as showing that we'd killed cavemen with herpes.

But that's not true! We actually just interbred and folded all their best genes into our own species, slowly spreading to become the dominant *Homo* strain in the process. So, really, you *could* say that we basically boned Neanderthals into oblivion. But herpes had little to do with it.

⊀

Now that you know that we've had sexually transmitted infections since roughly around the time we first emerged from the primordial ooze, you might wonder what ancient humans had to say about these audacious afflictions. The answer is that they had *plenty* to say and very little idea what the heck they were talking about. Relatable, no? Without the help of microscopy or anything resembling modern epidemiology, all sexually transmitted diseases congealed into a single, crusty mass of genital secretions and embarrassing itches. This made it impossible to know exactly who was catching what and when.

There is some debate, for example, on whether the ancient Greeks suffered from syphilis.[8] Yes, bodily fluids were weeping out of genitals in a most unseemly fashion, even during these simpler times. But these sorts of symptoms all oozed into a single, hand-wavy diagnosis; the descriptions provided by Hippocratic contemporaries were vague enough to have fit several different diseases. Herpes, bug bites, chicken pox, an allergic reaction to your new toga detergent . . . Who's to say what's causing the latest disturbance in your undies?

Before those days things were even murkier. Please keep all hands and feet inside the vehicle at all times: we're about to rocket through several thousand years of smelly discharge.

The Mesopotamians, by creating the first written language around 3100 BC, also in a sense invented history as we know it. And thanks to those written accounts, we know they almost certainly had some sort of venereal disease (VD). Promiscuity is common in their writing—the *Epic of Gilgamesh*, the oldest known work of great literature, features a sex scene that lasts *for a dang fortnight*—but unfortunately for us, they weren't particularly overt in their descriptions of infections tied to doing the dirty. Scholars have found medical texts documenting urethral and vaginal discharge, possibly due to gonorrhea, chlamydia, or trichomoniasis, as well as potential records of herpes lesions.[9] But it's possible these symptoms were due to other causes, like schistosomiasis—a parasitic infection that can lead to all sorts of bloody oozings (and you didn't even get to have sex while catching it, since the worms that cause it burrow directly into your skin while you stand in contaminated water).

The ancient Egyptians left similar signs. We know their doctors treated STI-esque symptoms, but we don't have any texts directly tying those conditions to sexual acts.[10] They called them "secret diseases." But whether this secrecy was due to stigmatization or simply the fact that they physically hid beneath your clothes remains unclear. After all, nuance is hard to read in tweets from five minutes ago, so don't be too hard on archaeologists for not knowing exactly what a crummy scrap of papyrus might be trying to imply about an ancient case of the clap.

The Hebrews got a little closer to spelling it out for us. The Book of Leviticus recounts a gonorrhea-like disease, which made priests advise patients to stay apart from their partners for a week (though this was kind of standard procedure for anything that made you "unclean," including menstruation, soooooooo). The Book of Numbers, meanwhile, features a story about Moses sentencing thousands of enslaved women to death to stop the spread of a new contagion among his soldiers (cool).

Thank goodness, then, for the Greeks, who were incredibly detailed in their records of all things medical. Herpes got its name during this era, being so christened for the "creeping" lesions that could cause tingling sensations as they erupted on one's skin.

But if these early physicians knew the skin diseases they studied were related to sex, they didn't seem to feel the need to make a big stink about it. Treatments for such ailments did not include abstinence. Instead, you might be told to sleep on a firmer bed or burn your little lesions with hot metal to keep them from spreading. If you showed up at the doctor's office with fiery discharge during Hippocrates's day, chances are good he'd tell you your humors were out of whack and give you some suggestions for better balancing your biles and phlegms. In fact, some medical texts describe prescribing more frequent sexual activity to *cure* various conditions that, in hindsight, were probably sexually transmitted. Whoops!

Moral hubbub over sexy skin conditions seems to have started with the Romans, otherwise known as the meathead off-brand Greeks. (I'm Italian! I can say this! It's true.) Their satirical poems started to crack jokes about diseases that were more common in men who had sex with men. And some Roman physicians praised virginity as the period of peak health, when various maladies were less likely to occur. In the early AD years, Pliny the Younger—the also-ran nephew and namesake of the man credited with inventing the concept of an encyclopedia—wrote down tales of a newly married couple who sought death together due to the inexplicable emergence of sores on the husband's genitals. This may be the first recorded instance of someone claiming to have passed an Immaculate Infection to their partner, but it certainly wasn't the last.

At first glance, you might think people had started to really crack the code on the whole STI thing by 1162, when the parliament held at Winchester reportedly banned women with "burning disease" from entering "stews," or brothels. Unfortunately for the workers and customers of twelfth-century stews, though, no parliament was held at that time in history. A fifteenth-century bishop actually drew up the ordinance—he just backdated it to add gravitas.[11]

As for actual twelfth-century STI know-how, the German abbess Hildegard of Bingen had a few thoughts to share on the subject. Bingen, also known as Sybil of the Rhine, was ahead of her time in a few ways. She composed music still performed today, studied astronomy, and once got an archbishop to move her and her nuns to their own monastery by claiming a crippling illness overtook her each time he refused. But her most dogged and best-documented pursuits were in botany and medicine.

The twelfth century wasn't exactly a progressive time in western Europe. Consequently, Bingen's writings are accordingly veiled in poetic language and laced with misogyny. For example, instead of noting calculations she may have made about the stars and seasons, she mused on the miraculous nature of space and our planet's cycles. When she wrote an allegory depicting the Catholic Church's corrupt officials as monsters, she couched the whole thing in a claim that it had come to her in a vision from heaven. And when she described, in uncanny detail, the female orgasm, she used language so flowery and detached that it's doubtful any of her male superiors questioned where and how a woman married to God had learned this.

Bingen also wrote of a form of curable "leprosy" that only struck "lustful men." That leprosy was quite possibly our good friend syphilis. And it seems pretty clear that the infection ushered in our first documented age of anti-STI sentiment.

Of course, all of this came before germ theory: the understanding that microbes such as bacteria and viruses cause disease, which didn't catch on until the nineteenth century. So even once scholars fingered sex as the culprit, the mechanism remained unclear. Some thought it was simply a question of literal filth on the genitals, while others blamed the wrath of the Lord or some kind of vaginal venom (yes, really).

This means that though they understood the infection was *transmitted* by sex, they didn't understand how exactly it was generated. As late as 1690, the paper "A New Method of Curing the French-Pox" reported that if a female virgin "kept company" with half a dozen young men, one of the fornicating parties would eventually—spontaneously—get the pox and pass it on to the rest. This, the physicians wrote, was due to the fermentation of mixed semen in the womb. This is, I'm sure we can all agree, *absolutely a thousand times more gross than how any actual STI works.*

Even as scholars continued to blame the transmission of syphilis and other STIs on everything from the sharing of clothing to the cross-contamination of eating utensils, the fact that it generally only struck people who were getting down with their bad selves (or, to be more precise, getting down with at least one other person, bad or otherwise) made it difficult to ignore the disease's carnal nature. With this knowledge came the first sparks of the ferocious stigma now attributed to STIs. But it also inspired a new and oh-so-slow-to-crest wave of medical research and a whole dang crew of physicians who were desperate to catch it.

Not literally, mind you; only a few scientists were willing to infect themselves in order to better understand STIs (more on that soon). Then again, large swaths of Europe would soon find they had little chance of avoiding them.

A POX UPON YOUR HOUSE

Given the tendency for writers like Aristotle to speak about sexual infections in such vague, easily confusing terms, just how and where our favorite squiggle-shaped microbe (yes, really, look it up) originated is impossible to pinpoint with any accuracy. A pox is a pox the whole world over, after all. But we know one thing for gosh-darn certain: whatever syphilis had been up to for the first three hundred thousand or so years of *Homo sapiens*'s history, it really began insisting upon itself during the fifteenth century, when King Charles VIII of France led an army of some fifty thousand men into Italy.[12]

"On their flippant way through Italy, the French carelessly picked up Genoa, Naples and syphilis," Voltaire would write more than two centuries later. "Then they were thrown out and deprived of Naples and Genoa. But they did not lose everything—syphilis went with them." The mischievous microbe had made its merry way to Africa and Asia by 1520. In what you might consider the sixteenth-century equivalent of "I got it from sitting on a toilet seat," various nations named the disease based on which country they decided it must have come from.

The French called it the disease of Naples or Spain, English and Italians named it the French pox, Russians blamed the Poles, the Poles blamed Turkey, the Japanese called it the Chinese disease, and the Turkish blamed Christians of all sorts. And so on.

Jacques de Béthencourt disliked the term *morbus gallicus* (French disease). And so, in 1527, he suggested that due to its sexually contagious nature—called out by an Italian surgeon in 1514 and widely accepted from that time forward—it be named instead for its connection to "illicit love." He suggested "malady of Venus," or *morbus venerus*, in more academic terms. This phrase morphed into "venereal disease." By the 1700s folks shifted from references to the *insert xenophobia of choice* pox to monikers such as the "bane of Venus,"[13] forcing individual patients to carry the stigma previously shouldered by entire foreign nations.

The word was out: you got it by boning. And boy, were people boning a lot. By the time Voltaire was smarmily writing on the subject, around 8 percent of adults in the British city of Chester had syphilis.[14] For reference, in 2016, health officials panicked over "epidemic levels" of syphilis when the City of New York reported some two thousand cases, or 22.7 infections per one hundred thousand residents, which comes out to an infection rate of around .02 percent. Chester! Wow.

It's no surprise that syphilis made such a mark on the historical records of this era. It wasn't just spreading at an alarming rate; it was also an incredibly alarming disease. For the first half century or so of its jaunt through Europe and beyond, the illness that resulted from the bacterium *Treponema pallidum* was much more virulent than it is today, probably because people had less immunity to it.

These days, the initial symptoms of syphilis generally include painless sores and itch-less rashes, though if left untreated the disease can eventually target your nerves, brain, heart, and eyes to cause debilitating and even fatal complications. It's also often fatal if passed to newborns during childbirth. In other words, it's not like syphilis is no big deal in modern times. But historical texts describe something more immediately devastating, with complaints like bone aches, excruciating pain, weeping sores, and organ failure.

Even once the disease turned from epidemic to endemic, syphilis patients living in a pre-antibiotic world were in for a world of hurt. Don't worry, though. The suffering of these countless poor souls was not in vain. We have their agonizing deaths to thank for one of the most hilarious pieces of art in human history.

SALACIOUS ARTICHOKES AND BESOTTED BOYS

Sure, you think you're a syphilis enthusiast. I get it. Who doesn't? But have you ever written an epic poem—like, epic in the style of Ovid, not just *totally epic, dude*—about the fabulous French pox? Such a thing sounds too good to be true, to be sure, but I assure you it exists. In fact, the eighty-four-page poem written in 1530 was popular enough to inspire a lovingly—and I mean *lovingly*—executed translation in 1686.

Girolamo Fracastoro (along with his seventeenth-century translator-cum-fanboy Nahum Tate, who sported some truly spectacular wigs) saw syphilis as a plague so monumentally historic as to warrant "a poetical history" on the subject. The Italian Fracastoro was well versed in the natural sciences of the day ("He was so well educated by his Father that he gave early proofs of a great Genius, so that in his childhood all men conceived hopes of an extraordinary Man," according to the totally-chill-about-this-subject Tate). Even so, Fracastoro's poetical musings were more art than medicine. He wrote another, more straightforward text on syphilis that provided useful information, at least relative to the available knowledge of the day; the poetical history was what he did for fun. I am certain many of us can readily relate.

And, well, the results of this artistic endeavor were spectacular (or at least spectacularly bad). One could quite easily write a poem in praise of Tate's translation of Fracastoro's poem, not to mention of the original poem itself. Tate seems to think Fracastoro produced a truly great work of art: he even points out in his foreword that the original poet survived a lightning strike that killed his mother as she held him, as a means of highlighting the providence of his existence. And Tate produced such lyrical translations as a description of artichokes as "salacious."

My appreciation of this great work was only magnified by the lucky circumstances under which I perused its lyricism. I had the good fortune of breathlessly turning the pages of one of the original copies of the book in the New York Academy of Medicine's rare book room (fun fact: anyone can make an appointment online to sit and read really old books about diseased penises there). Seeing the meticulously hand-inked text of Tate's work really hit home that *someone spent their time writing a really long poem about syphilis that features delights including but not limited to nymphs and a diss on cucumbers.*

Truly, reader, I recommend it. Here's just a taste:

In that dire Season this Disease was bred,
That thus o'er all our tortur'd Limbs is spread:
Most universal from it Birth it grew,
And none have since escap'd or very few;
Sent from above to scourge that vicious Age,
And chiefly by incens'd Apollo's Rage,
For which these annual Rites were first ordain'd,
Whereof this firm Tradition is retain'd.
A Shepherd once (distrust not ancient Fame)
Possest these Downs, and Syphilus his Name.
A thousand Heifers in these Vales he fed,
A thousand Ews to those fair Rivers led:
For King Alcithous he rais'd this Stock,
And shaded in the Covert of a Rock,
For now 'twas Solstice, and the Syrian Star
Increast the Heat and shot his Beams afar;
The Fields were burnt to ashes, and the Swain
Repair'd for shade to thickest Woods in vain.

It is perhaps surprising that a man of science would spend his time writing an eighty-four-page poem about syphilis. Perhaps it is even more surprising that another man of science would spend his time translating it a century later. But to understand the motivations of our salacious friend Fracastoro, we need only dip our toes into the known history of the disease.

Here's the thing, my salacious artichokes: the dudes were not wrong. Epidemiologically speaking, syphilis was—and is—*epic as all heck.*

Europe had a fever, and the only prescription was . . . more mercury.

ONE NIGHT WITH VENUS AND A
LIFETIME WITH MERCURY

Mercury had such staying power as an STI treatment—specifically for syphilis, but, given the general confusion about how many sexually transmitted infections existed and what they looked like, let's be real, probably for every other STI too—that it reportedly gave rise to the adage "a night with Venus and a lifetime with mercury."[15] This is very clever. We like this.

But we do not like the use of mercury for curing STIs. During the peak syphilitic craze of the 1500s, the general course of treatment was to cover yourself in mercury ointments and sweat and salivate as much as possible to purge yourself of bad vibes.[16] This could go on for days, weeks, or months. And it was considered an improvement over the option of drinking mercury elixirs—which were increasingly recognized as toxic—even though letting mercury seep into your skin and vaporize into your lungs is absolutely, definitely, decidedly still toxic. You might also ingest the sap of the guaiacum plant, which had been brought over from the New World by Spaniards and was generally considered a panacea. It did not cure syphilis.

Mercury would maintain its status as the ultimate syphilis smasher until the turn of the twentieth century, when new medical advances offered a better option: arsenic. And malaria.

"Pyrotherapy" is an extremely cool word. And it doesn't even require you to literally set your patient on fire. No, you just have to give them an artificial fever. The notion is that, just as a genuine fever can help you recover faster if it doesn't kill you first, a sufficient bump in temperature driven by external forces can help get you back on your feet. (The mechanism by which fevers help cure us, by the way, as well as how much good they do for various diseases, is an ongoing research topic, but it's possible that higher temps create a less hospitable environment for certain pathogens and also that warmth speeds up the cells that drive our immune responses.)[17]

In the 1880s, Austrian psychiatrist Julius Wagner-Jauregg set out to see how pyrotherapy could help various mental health conditions after seeing a patient with psychosis seem to recover from it after a nasty bout of bacterial infection. Then, in 1917, he somehow got his hands on a malaria patient and helped himself to the man's blood before curing him with quinine. That blood went straight into the arms of several of Wagner-Jauregg's neurosyphilis patients—individuals suffering cognitive decline due to the final stage of the disease, when it attacks organs like the brain—who were allowed to achieve high malarial fevers for some time before getting quinine.[18]

Modern analysis suggests this technique really was quite effective, though it carried a nonzero risk of death due to the fever itself and wasn't 100 percent successful. It also, at least in some cases, drove hospitals to infect unwitting healthy people with malaria so they would have a safe reservoir to use for treatment.[19]

But it was good enough in the face of neurosyphilis to win Wagner-Jauregg (fun fact: a Nazi sympathizer!) a Nobel Prize in 1927. And its effectiveness persisted until the 1940s, when penicillin was widely adopted as a much safer and less ethically icky option.

WHENCE CAME THE CLAP?

Gonorrhea is one of the world's oldest extant STIs.[20] Google will tell you that it is called "the clap" because of a horrific treatment, whereby penises were "clapped" between the pages of a large book as a means of forcing the pus out. We don't actually know for sure if this is true. Some experts suggest the name somehow refers to the sensation an infected person feels while peeing; others say it comes from the fact that French brothels were known as *clapiers*, or rabbit huts.

In any case, even if no one was clapping books onto penises as a regular practice, gonorrhea had some questionable treatments.[21] There are reports of sixteenth-century men having mercury (our old pal) injected into their urethras. And even an eighteenth-century European treatment could range from a bland fluid diet to something called urethral lavage or irrigation—wherein male patients were catheterized and then had scalding hot water flushed through their willies for several days in a row. According to historians, the success of the treatment was considered to be directly proportional to the agony experienced by the infected person.

In addition to the usual poisons, purgatives, and sweating regimens you'll find in the treatment of any STI throughout history, Europeans in the 1800s also turned to two botanical cures. There was *Piper cubeba*, an Indonesian plant used by Arab physicians and alchemists since antiquity and said to taste like something between allspice and black pepper, and there was copaiba, the extract of a South American tree. This was a hot ticket: in the first ten months of 1859, some 118,396 pounds of copaiba resin were imported into Great Britain, and it's thought that the stuff was almost exclusively used to treat gonorrhea.[22]

But did it work? Freeman Bumstead, an American physician considered one of the leading authorities on VD during the Civil War, reported mixed results. The plants, he said, "are of undoubted efficacy in the treatment of many cases of gonorrhea, but in others they utterly fail; nor have we any means of distinguishing these two classes of cases beforehand . . . [T]hey are by no means indispensable in the treatment of every case of gonorrhea."

So, you know, 50 percent of the time they work 100 percent of the time.

Bumstead may have been devoted to understanding sexually transmitted infections. But he was far from the most dogged investigator of their origins.

PRICKING PRICKS FOR SCIENCE

Like many in the '60s, John Hunter was big on the concept of sexual experimentation. But given that these were the 1760s, that experimentation wasn't so much of the psychedelic or free-love variety: it mostly involved poking people with needles covered in pus.

Scientists were just beginning to understand how disease moved from point A to point B, and Hunter was at the head of the pack. Most of his colleagues saw the angry, oozing infections that festered in soldiers wounded in the Seven Years' War to be a natural—nay, a healthy and desirable—step in the healing process. But Hunter understood that this rank inflammation was worth avoiding. He started a trend of leaving bullets inside the body as long as they posed no immediate threat, thereby avoiding unnecessary surgery amid the unhygienic theater of the battlefield. And yes, keeping a hunk of lead inside you really was preferable to surgery. Extraction, after all, would entail widening the hole in your sad meat sack (presumably with some extremely not-sterile tools) so a surgeon smothered head to toe in other dudes' viscera could get his fingers inside you. In what eighteenth-century physicians considered a real plot twist, Hunter's patients survived because he *didn't* cut them open.

But Hunter, a maverick surgeon and the most popular jack of his trade in London, wasn't just an early advocate of basic wound hygiene. He also had one very particular bone to pick in his field of study. And, as a result, he eventually also had syphilis.[23]

Like Sleeping Beauty reaching all awestruck for a shining spindle, Hunter had (allegedly) pricked his own—well, you get the idea—with a needle doused in the unfortunate emissions of someone with gonorrhea. This was on purpose. Sort of. The syphilis was an accident, but Hunter was pretty pleased about it. Yes, this is confusing. Most of medical history is.

Here's the deal: Back then, our grasp of microscopy was limited to the identification of wriggling blobs on a slide. So it was reasonable to argue all the live-long day about whether various symptoms came from one disease, three diseases, ten diseases, whatever. If a whole family is sneezing and coughing, but only one of them has a fever, modern medical know-how tells us it's possible, or even likely, that they all have the same bug. We know that one pathogen can often affect different individuals to different degrees and in slightly different ways. But such a logical leap would have been more controversial in Hunter's time. And this same level of confusion persisted among the care and treatment of STIs.

Hunter believed that syphilis and gonorrhea were caused by the same microbe (spoiler: they are not). Because he'd seen gonorrhea clear up on its own, he argued against the aggressive treatment of syphilis, at least in some cases. But to support his argument, he had to show the diseases were one and the same. The best way to do this, any eighteenth-century surgeon would no doubt tell you, was to infect someone using the pus of a gonorrhea patient.[24] Hunter did this to a few people, possibly including himself (Are you confused? So is the historical literature!) and was triumphant when said study subjects started exhibiting syphilis symptoms.

It apparently never occurred to him that patients he sourced his experimental pus from might be veritably riddled with infections of all sorts, leading Hunter to inoculate his test subjects with a good old grab bag of bacterial invaders. The human body contains multitudes, and there's room for more than one venereal disease on this corporeal party bus.

Many modern scholars argue that Hunter *actually* performed these experiments on random patients. Indeed, infecting unsuspecting individuals with various diseases, syphilis included (and, sometimes, a malaria chaser to try to cure the syphilis, which is an extremely old-woman-who-swallowed-a-fly sort of situation), was par for the course until very recently.[25]

As late as the 1970s, American researchers with the CDC lied to *hundreds* of black men about their health while withholding penicillin in order to watch the progression of their syphilis. The Tuskegee Syphilis Study is a shameful moment in medical history, but it's not uniquely horrific. On the whole, stabbing one's own groin with a dirty needle is, frankly, a lot less shocking and a lot saner than experiments carried out on the disenfranchised by the US government. Never mind the sorts of ad hoc studies conducted on enslaved people, prisoners, the poor, and people with disabilities throughout all of human history.

Consider the unsettling fact that Hunter bought and displayed the bones of an Irish giant who had asked to be buried at sea.[26] Resting forever beneath the waves, to be clear, is basically the opposite of spending eternity sitting on some English dude's bookshelf. So the good doctor wasn't exactly a paragon of respect for bodily autonomy.

In summary, Hunter was absolutely the sort of guy who would stab some random poor person with a dirty needle. Most physicians did so with shocking frequency until very recently.

But whether Hunter was a rare hero willing to put his own junk on the line or merely the kind of unethical jerkwad responsible for most medical research in our species' history, it seems possible that he *wanted* the public to believe he'd contracted syphilis himself. The story that he did so comes from notes taken by his own students, his own lectures, and one of the editors of his scientific papers. Imagine a world in which surgeons spoke proudly of their syphilitic lesions, the societal stigma of which paled in comparison to the great triumph of pushing scientific knowledge forward with the power of their own conviction. Sure, Hunter's conclusions were very, super, extremely incorrect. But he sure as heck was proud of how he'd made them.

Given the public's reaction when Aaron Traywick injected himself with genetically engineered herpes live at a biotech conference in 2018, it's safe to say that Hunter's supposed self-experimentation would be seen as just as shocking, if not more so, today. Traywick's decision to try out an experimental HSV vaccine went all the wrong sorts of viral. Media outlets painted him as a renegade madman, with many commenters on sites like YouTube and Reddit snickering over what sort of stuff he must be getting up to, to want such a shot in the first place. And when Traywick died in an unrelated accident that same year, clickbait articles giddily implied his death was related to his experimentation with herpes.

Don't get me wrong: I'm not advocating for the replacement of rigorous medical trials with showy self-experimentation on the world stage. But the fact that exposing yourself to an STI in the name of science has seemingly become *more* unseemly since the 1700s should, perhaps, give us all pause.

IF IT QUACKS LIKE A DUCK

For much of history, STI treatment was expensive, painful, and middlingly effective. So, just as people with chronic discomforts and even full-blown medical conditions are often wooed by wellness influencers today, you won't be surprised to learn that folks with oozing sores and itchy bums were often taken in by quackery in the era before antibiotics and antivirals. After all, a doctor or surgeon was just going to price-gouge you for mercury—a cure you'd heard was woefully unpleasant. Why not try something cheaper that was sold to you with a smile?

In Victorian London, quacks did more than just peddle snake oil from door to door or in the post. They also provided everyday folks with some of the only sex education they could access.

Joseph Kahn's anatomical museum started out as a classy place. In 1851, having emigrated from the part of Europe that was still ping-ponging between being French and being German, he put down roots at 315 Oxford Street in London. He and his family opened a Museum of Anatomy and Pathology. This was a common-enough concept, starting in the 1700s, where public visitors could pay to see wax models of the human body in various states of dissection, along with assorted specimens preserved in alcohol and, in the case of Kahn's museum at least, microscope slides. Historian A. W. Bates notes that these sorts of exhibits, which had been quite elaborate in the eighteenth century before sliding out of fashion, were booming again by the time Kahn came along.

This may have been largely due to the public trial of William Burke and William Hare, who in 1828 killed at least sixteen people and sold them to the University of Edinburgh's anatomy department for dissection.[27] The procurement of cadavers for study had always been a contentious issue, which could be and definitely has been the subject of its own whole book. But, suffice it to say, those dastardly Williams got Victorian folks jazzed about viscera again.

This was no circus sideshow or tawdry attraction. Kahn won the approval of the British medical journal the *Lancet*, which was known for sniffing out quacks at the time, especially in regard to his collection of embryos and accompanying explanation of the processes of fertilization and development.[28] But he also had a room—just for medical professionals, mind you—where one could gaze upon vivid reproductions of genitals in the throes of venereal disease. Given that most doctors would have already seen these symptoms in person, the exhibits for medical professionals were likely simple enough to access with a wink and a nod in lieu of clinical credentials. Indeed, Kahn on several occasions appeased *Lancet* critics by cutting down on the number of exhibits that he claimed to allow visiting ladies to observe, as if he and his staff would actually pack things up in preparation for a skirt to enter. He did, however, openly allow women with "professional interest"—nurses and midwives—into the gorier parts of the museum.

But that was before Kahn fell on hard times. That was before he met the Jordans.

The Jordan family were the sort of quacks the *Lancet* loved to go after. Under the name Perry and Co, they claimed to be self-taught medics, selling various cures for VD by post, which they advertised in public urinals. Kahn initially hosted lectures that vehemently dismissed Perry and Co's products but was soon bought off by way of a large order for wax models. Once they got to talking, Kahn pivoted his museum to have an STI focus—and to help peddle Perry and Co's cures.

I want to pause to imagine how this could have been a great thing. For an everyday Victorian Londoner, the ability to see wax models of STI symptoms, learn about treatment options, and ask for prevention advice from medical experts would have been completely game-changing. Kahn's place could have been like sex-ed Disney World, which is a concept I'd love to bring back.

Unfortunately, cash is king. And Kahn's intentions were less noble than putting genuine knowledge into the hands of the working class. Instead, the establishment became little more than a place to sell Perry and Co's remedies, including to people with no reason to think they had STIs anyway. I haven't been able to track down detailed descriptions of what they were selling, but the history of medicine tells us it was probably mostly water.

By the mid-1850s, several medical journals that had endeavored to expose "the gang of vile Jew Quacks"—which I do not love, for reasons I hope are obvious—added the once-admired

Kahn to their hit list. His downfall was assured in 1857, when he went on trial for extorting a client who asked for his money back. He allegedly threatened to publicly accuse the plaintiff of masturbation in retaliation. "Oh!" the judge is said to have exclaimed in court, "even if it were true, it would be a monstrous thing for a medical man to assert." The *British Medical Journal* offered that no melodramatic novel had ever invented a "more odious character" than Kahn. This seems a bit much, but okay.

Kahn soon vanished into obscurity, but the museum stayed open until the 1870s under the Jordan family's management. Intriguingly, the Obscene Publications Act of 1857—which we'll talk more about in our porn chapter—greased the wheels for the place to be shut down. England had no way of directly targeting people who sold quack cures; the Medical Act of 1858 only regulated qualified practitioners who registered as such. All the law could do about people like the Jordan family was shout "NOT A DOCTOR" at them, which potential clients managed to ignore in the same way they ignore similar warnings about unregulated supplements and homeopathic nonsense today. But their advertising and experiential marketing in the form of a museum could be and indeed was, eventually, dinged as "obscene." And that spelled the end of this egalitarian exchange of medical knowledge (and, you know, quackery).

NO LAUGHING MATTER

I don't have any silly stories to tell you about HIV. Human immunodeficiency virus is a reminder that while we shouldn't be *more* afraid of STIs than we are of other infections, we shouldn't forget to be wary of them.

Look at it this way: You're not going to spend the rest of your life inside just because you know another virus like SARS-CoV-2 could emerge at any time and probably will at some point in the future. But you can still recognize that there are steps we should all take as a society to keep the COVID-19 pandemic from happening all over again. People should have access to paid sick leave so they're not pressured to sneeze all over the subway. Governments should invest in shoring up each nation's ability to produce personal protective equipment at a moment's notice. Motherfuckers should wash their hands. And so on.

If the COVID-19 pandemic teaches us anything, I hope it will teach us that we can recognize the potential danger of a pathogen—and take steps to avoid catching and spreading it—without focusing on outsize fear or stigmatizing people who are most at risk. That's probably too much to hope for, since I don't think we did a great job of learning that lesson even specifically for SARS-CoV-2.

But if you ate at an outdoor restaurant patio while unvaccinated, you shouldn't say you'd never have sex with someone with HIV. To be clear, you can always decide not to have sex with someone. Maybe this person has HIV and is *also* a raging asshole, whatever. It's your body and your life. I'm just saying that if that's your *only* reason, your logic may be fundamentally flawed.

Because here's something about HIV that really makes me smile. People who contract HIV have to take their antiretrovirals to make sure they don't get seriously ill. But if they take them regularly, they can often get their HIV cell count down so low that the virus is functionally (though not literally) undetectable in a blood test. Research has proven, definitively, that people who have undetectable HIV loads in their blood can't transmit the virus.[29] It is increasingly possible for a person with HIV to get confirmation from their health-care team that they are not at risk of transmitting HIV to their partners. Meanwhile, people who don't have HIV but are at risk of contracting it from their partners can take drugs like Truvada that nearly eliminate that risk. Along with responsible condom use, drugs like these *really and honestly* have the power to make HIV transmission just . . . stop happening.

We may not yet have a simple and complete *cure* for HIV. But between these antiretrovirals and general safer sex practices like using condoms and employing plenty of lubrication to avoid skin tearing during sex, we have the tools we need for HIV to be a chronic condition instead of a death sentence and to keep it from continuing to spread. Now, the equitable distribution and implementation of those tools is a whole different hurdle. And we don't know what other STIs might evolve in the future. But don't be afraid. Be careful, be honest, and be kind—not afraid.

Okay, I actually do kind of have one funny story about HIV/ AIDS. Here we go. So, in the early 1980s, no one knew what the strange "cancer" that seemed to target gay men actually was. That led to a lot of horrible stigma and preposterous theorizing and, according to at least one article in the *Los Angeles Times*, a decline in hot-tub sales.[30] This isn't the funny part.

What's funny is that in 1985, a study linked Kaposi's sarcoma, one of the most common infections in AIDS patients, to the use of poppers, or amyl nitrites. Originally rolled out as a Victorian-era heart medicine, these vasodilators widen your blood vessels when you inhale them. This lowers your blood pressure and increases blood flow, in addition to relaxing involuntary muscles. The combined result is a brief euphoric head rush, a warm, tingly feeling, and orifices that are, shall we say, ready and eager to receive. Poppers are not great for your brain, probably, if you do them a lot, and they're definitely bad if you drink them, which really is just a reminder that you should not do drugs you haven't, like, googled. (Seriously, there keep being case reports of people drinking poppers, and this is not how *anyone who knows anything about poppers will tell you how to take them*, so I must once again ask if the heterosexuals could kindly desist.)

Anyway. There's *no* mechanism by which using poppers could give you or make you more likely to get HIV. That didn't stop a full-blown conspiracy theory from cropping up about Big Popper causing the AIDS epidemic.

In fact, as journalist Alex Schwartz reported for *Popular Science*, poppers might actually reduce the risk of HIV transmission during anal sex by loosening sphincters and reducing the risk of bleeding.[31] News you can use!

A BUG WORTH CATCHING

You can catch a lot of dangerous stuff while having sex: urinary tract infections, STIs, feelings. But while inflamed urethras suck and emotions are pretty much never worth having, there's reason to believe that some sexually transmitted microbes can have serious upsides.

In 2015, biologists at the University of Texas outlined four documented instances of sexually transmitted infections that benefit their hosts.[32] Yes, four! And those are just the ones we already know about. Given that our research on the complexity of the world's microbiomes is practically in its infancy, it seems likely there are other fungi, bacteria, and viruses that do a body good and are most efficiently shuttled around by way of sex.

Aphids, for example, trade around a bug called *Hamiltonella defensa* by butting uglies. These microbes come in handy when parasitic wasps come around to lay eggs inside the aphids' bodies. Instead of bursting with parasitic larvae, aphids infected with *Hamiltonella defensa* can kill the grubs while they're still tiny.[33] *Hamiltonella defensa* is so useful, in fact, that it might be what drives aphids to have sex at all—the females are perfectly capable of reproducing asexually and do so most of the time. But if their own mothers don't pass *Hamiltonella defensa* down to them, they can only get the microbe for themselves and their own spawn by finding a male who's already infected. *Anopheles stephensi* mosquitoes make a similar exchange, with males gifting females bacteria of the genus *Asaia* during sex.[34] Research suggests that when females in turn pass this to their offspring, it shortens the larvae's necessary development time by as much as four days, perhaps by providing some kind of nutritional assistance. When certain fungi fuse to reproduce sexually, they can trade viruses that aid in growth and thermal resistance or produce toxins to kill off competitors.

Believe it or not, humans have at least one beneficial STI as well.

GB virus C (GBV-C) is a relative of hepatitis C that doesn't seem to cause any sort of disease in humans.[35] That makes it easy for us to keep passing around. Studies of donated blood in developed countries find it in 1 to 5 percent of donors, while as many as one-fifth of blood donors in developing countries tend to carry it. For most people, GBV-C probably floats around as a nonentity. It's just one more anonymous resident in the bustling metropolises that make up our microbiomes.

But for reasons unknown, GBV-C changes how people infected with HIV fare. Research suggests that having both makes a person 59 percent less likely to die of illness related to HIV.[36] There's also evidence that carrying GBV-C can improve outcomes for people with Ebola.

Now, please don't take this as an excuse to go raw dog the whole planet so you can collect nifty microbes like Pokémon. The same scientists who are tracking down GBV-C's exploits are eager to find ways of isolating the power of this and other beneficial microbes so we can acquire them in a safer fashion.

In the meantime, the benefits of using condoms, dental dams, and other accoutrements vastly outweigh the downsides of missing out on mysterious sexual microbes. The year 2019—the most recent with complete CDC data—saw the highest rates of STIs in the United States *ever* for the sixth year in a row. Chlamydia, gonorrhea, and syphilis were reported in 2.6 million instances, representing a one-third increase for the same trio from 2015.

But if you tend to think of STIs as oopy, goopy, skin-crawling afflictions, remember that they're no better or worse than any other category of microbe. Yes, they can absolutely mess you up. But that's not because you get them from screwing.

If anything, STIs are a lot *less* scary than other transmissible illnesses out there. Ebola? That'll sure catch you off guard. West Nile? Good luck avoiding every mosquito on the planet for the rest of time. You can get bit by a lone star tick and be allergic to meat for the rest of your life. You can catch a deadly case of the flu just by being in the wrong place at the wrong time.

But we know how to prevent STIs. Wear a condom! Use a dental dam! Get tested between partners! Get an HPV vaccine! Take Truvada if you or your partner is at risk of transmitting or catching HIV! Don't fuck people who aren't willing to do and talk about all these things!

Practicing safer sex won't guarantee you'll never catch an STI. Herpes, for instance, is spread via skin-to-skin contact, and I don't see gimp suits going mainstream anytime soon. And I can't promise that getting one will be no big deal. But if you're careful, you'll cut your risk way down. So few things in life are that simple. Take the easy win on this one.

Chapter Seven

HOW IS BABY FORMED?

· ·

In which semen coagulates, poisoned blood con-
geals, and tiny men float on the breeze to beget life.

· ·

DEARLY BELOVED, WE ARE GATHERED HERE TODAY TO TALK ABOUT THE
time everyone thought there were tiny little men folded up inside
every sperm—or maybe every egg. More on that in a minute.

While it's easy enough to scoff at the stupidity of scholars
of yore—especially male ones pontificating on how pregnancy
might work—it's understandable that the whole thing was so
long shrouded in mystery. We know that lots of ancient peoples
grasped the basic concept that heterosexual intercourse could
lead to a baby. One needs only to crack open the Old Testament
and take in all the tales of so-and-so lying with what's-her-face
and begetting someone-or-other to see proof. It makes sense that
further details remained obscure. For most of human history, the
only way to see what was going on in a human body was to cut it

open, and even then your findings would be limited to what you could see with your unaided eyeballs—not to mention tainted by virtue of the subject's current state of unaliveness. And, fortunately, not many scholars went around slicing pregnant people in half in the name of science. Even if they had the rare occasion to investigate the corpses of a late mother and fetus, what they observed about gestation couldn't tell them much about the mechanism of fetal creation. Until microscopes allowed us to see sperm and egg in action during the late seventeenth century, we were shooting blanks in the dark. And microscopes wouldn't actually become powerful enough to get a clear picture of the process for many years more.

Consider the twelfth-century scholar Hildegard of Bingen (remember her from those sexual leprosies?), who held that semen was just blood that had been curdled and poisoned by mankind's fall from grace in the Garden of Eden. Only the female womb could warm this noxious witch's brew and make it suitable to beget life. The strength of the sperm determined the sex of the child, Hilde said, while the passion of the parental match determined an offspring's disposition. In the worst case—a weak-spermed man with no love for his wife—parents could expect a bitter daughter. Further back in history, Aristotle had been so convinced of the progenitive powers of sperm that he referred to females as "infertile males."

As we can see, getting conception right was pretty hard. And we're about to go on a wild ride of ridiculous ways our forebears thought they'd conceive us: everything from mixing male and female "semen" (gross) to needing to make women orgasm first (good idea, but no go). But before we let the mockery commence, we need to take a good, hard look at ourselves. Because—and I don't mean to offend you, but it's my job to report the facts—*you are also* probably wrong about how pregnancy works.

Don't feel bad; this isn't your fault. The mechanisms of conception have been purposefully obscured by smoke and mirrors and conservative politics for as long as anyone reading this has been alive.

For example, you probably think that most people are pregnant for about forty weeks before giving birth. As in, there's a fetus (or, to be more precise, a zygote and then an embryo and then a fetus) growing in a uterus for forty weeks before it makes its exit, unless things go awry.

But when a pregnancy hits the so-called forty-week mark and acquires full-term status, the actual pregnancy has only actually existed for thirty-eight weeks. And it's actually a bit of a stretch to even say the pregnancy has lasted for *that* long.

I know, I know. This is all very silly. But it's important that you understand how pregnancy actually happens.

If you're a person who ovulates, one of your ovaries typically expels an egg about two weeks before your period is scheduled to arrive, give or take a day or two. This is, by the way, less of a "release" and more of an "ovum busting out of an ovarian follicle like a race car driving through a brick wall." All of this can only occur if your brain's pituitary gland releases the right hormones to tell your ovary to produce a follicle, which is a sort of fluid-filled cyst that secretes the hormone estrogen to toughen up the walls of your uterus, and if one of those follicles keeps growing for almost two weeks. Once it's mature, the resulting surge of estrogen tells your pituitary gland to spurt out some luteinizing hormone, which makes the follicle grow even more. Then the swelling follicle—which by now has started sending chemical signals to your fallopian tubes, whispering for them to move close and prepare for liftoff—bursts open to let the ovum seek its fortune.

In other words, this is already a lot more complicated than "sperm meets egg."

One of your two fallopian tubes should now be ready to run interference; it caresses the ruptured follicle with fingerlike protrusions called fimbriae to sweep the egg out of your open abdominal cavity (yes, it's just . . . out there) and into the fallopian tube. The muscular tube then squeezes the egg toward the uterus, which it will reach in something like eight to ten days. But the ovum will only survive as a free agent for around twelve to twenty-four hours, so if it doesn't encounter sperm within that time, it will show up in the uterus DOA.

The sperm, meanwhile—tens of millions of them from even a single ejaculation—will thrive in the acidic cervical mucus produced during fertile periods, and a select few will be pushed into the fallopian tube with small uterine contractions. They can live for up to five days in a pinch, so it's possible for gametes to find their way to an egg even if sex happened days before ovulation. Driven by attractive chemicals released by the egg and aided by an increased number of hairlike structures called cilia that line the walls of the fallopian tube, some small number of the initial horde of sperm will make their way to the egg, often in a manner of minutes. During this process, hormones will trigger a stage called capacitation where the sperm fully matures; its tail gets a bit more freewheeling to help it swim faster, and the membrane on its head destabilizes in anticipation of combining with the ovum it seeks.

Like ovulation, the act of fertilization is probably more brutal than you've been taught. The egg and sperm don't just embrace and become one. First, the sperm releases an enzyme to dissolve the "cloud" of cells surrounding the egg called the cumulus oophorus. Then, sperm—sometimes more than one—fuse to the egg's outer membrane and start to digest it. This membrane turns into an impenetrable shield once a sperm cell actually makes it into the egg's innards, ensuring nothing will interrupt it as it wiggles toward the egg's nucleus. Once there, the chromosomes of the sperm and egg will finally merge into one cell—a zygote—and start dividing and replicating.

Okay, now we can finally get back to the confusing timing issue we started out with. See, the creation of the zygote doesn't really constitute a pregnancy yet. First it has to make it to the uterus, which takes almost a week or even close to two. If it gets there and has kept dividing and growing the whole time, it's known as a blastocyst. You're still not pregnant.

If everything has gone according to plan, hormonally and physically, the uterus will now release chemicals that degrade the blastocyst's protective outer membrane, and the growing cluster of cells will anchor itself firmly into the uterine wall. Now, finally, about two weeks after ovulation, you've got yourself an embryo. This is no small feat: about half of all fertilized eggs will fail to make it to this point. Half!

Here's the rub. We don't start counting the forty weeks of pregnancy at the moment the would-be embryo successfully latches onto the uterine wall. We don't even start counting at the moment that a sperm cell drills its way into the egg to make a zygote. We start counting on the first day of your last normal menstrual cycle. That's two weeks before conception, and possibly nearly a *month* before there's actually something viable growing in your uterus.

This is really stupid.

But it's arbitrary, right? Who cares? *You* should care, because it has a huge influence on what options are available to you should you become pregnant accidentally.

A rise in so-called "heartbeat" bills aims to limit abortion after six weeks, ostensibly because this is when a fetal heartbeat can be detected on an ultrasound. Quick side note here: what a doctor might pick up at six weeks is *not* a heartbeat. There's a reason that medical professionals don't actually switch to calling an "embryo" a "fetus" until around eleven or twelve weeks; that's the point at which organs have formed, and all the pregnancy really has to do is foster growth, not the creation of new body parts. That's why about one in ten pregnancies fail in the first trimester and why about three-fourths of all miscarriages occur during this period of crucial development. The embryo spends these weeks turning random cells into a whole-ass future being.

At six weeks (which is, once again, more like four weeks or two weeks of actual pregnancy, depending on how you look at it), the four-millimeter-long embryo consists only of clusters of cells that may *one day* form organs like a heart or a brain. Modern ultrasound technology is now sensitive enough to catch the electrical activity of these cells, which is a sort of rhythmic pulsing.[1] That's the signal of sanctified life that heartbeat bills refer to.

Besides being based on a falsehood, these bills present a disingenuous sense that people still have a reasonable amount of time to terminate a pregnancy. Surely folks in want of an abortion can simply get it done in six weeks, no? Well, that's where the weird way we track pregnancy progression turns into a big problem.

A person is in their second week of "pregnancy" by the time conception occurs (talk about revisionist history). Pregnancy tests take at least another week to work and are most reliable another week after that. So now, if you're not eagerly taking high-end pregnancy tests—something people who aren't trying to get pregnant are understandably unlikely to add into their daily routines—you're "four weeks" along before you have any reasonable chance of knowing you're expecting. If you're truly blindsided (as many people who get pregnant unintentionally are—the point being that it's not intentional), it could take another week or so before your period is late, which is when most people will be prompted to take a test.

That means it is *100 percent understandable if you're "five weeks pregnant" when you realize you're pregnant.* And this is especially understandable given that you'll only have had a zygote growing in your uterus for about a week or so and probably only had the sex that resulted in this pregnancy a couple of weeks before.

So, if you're lucky, you have one week—maybe two—to find a clinic, make an appointment, and get an abortion. In states where these laws exist, clinics are often far enough apart to require hours of travel; people seeking abortions are often forced to undergo an ultrasound and wait twenty-four hours before actually getting a termination, and such services may not be covered by insurance, if individuals have coverage at all. When you factor in the logistical headache of getting time off work, figuring out transportation, organizing accommodation for a forced overnight stay, arranging care for pets, kids, and any other dependents, and finding the money to pay for it all, those one or two weeks swirl down the drain pretty quickly. And that's assuming you *have* a week or two. Because menstrual cycles can be thrown off by illnesses, medications, stress, and other factors, it's completely understandable for someone not to take a pregnancy test until a week or two after they expected their period to start.

That's why the science and history of pregnancy matter—and why we don't have much right to laugh at the "science" of pregnancy that came before. But we'll do it anyway.

A SEMINAL DISCOVERY

Now that we know a bit more about how hard it is to even understand conception *today*, perhaps we can better understand how hard it was for those humans who didn't have microscopes and the internet at their disposal.

In the 1660s and 1670s, respectively, Jan Swammerdam and Marcello Malpighi produced experimental results that would convince fellow scientists that all living things must exist in teeny-tiny form inside their mother's eggs. In comparison, existing alternative theories were downright nonsensical. Many naturalists favored spontaneous generation for simpler animals: a pile of dirty socks would beget mice, naturally. Even complex humans were thought to stem from epigenesis, a process whereby male and female fluids just kind of gummed together until something started to solidify. This notion required the intervention of some higher power to turn sex goop into human flesh, because, you know, nobody was talking about stem cells in the 1600s. So even though it was technically sort of right, it was also all sorts of horribly wrong.

Enter preformationism, based very logically on the observation of metamorphosing insects and growing chicken eggs. If tiny, fully formed chicks eventually grew and hatched from eggs, why couldn't humans do the same? In its heyday, so-called ovism, a subgenre of preformationism, posited that all life had existed at the moment of creation, with future generations tucked inside each ovum, Russian-nesting-doll style.[2] What a complex, whimsical, logical, nonsensical, poetical theory for the existence of all things. Preformationism gave humankind a brief moment of believing that we'd all seen the dawn of history, tucked as we were within the gonads of our mother's mother's mother's, etc.

Preformationism was in many ways progressive for its pro-ovum bent. Indeed, a man's ejaculation was thought to merely trigger some sort of explosive growth of the homunculus, which had been tucked sleeping for generations inside the female egg. Alas, this exultation of the female gamete was not due to some sudden feminist Gaia woo-woo-ing on the part of contemporary scientists, but only because eggs large enough to see without microscopes existed in nature. Sperm had only been detected as a wriggling mass, which left scientists assuming it was a parasitic worm that happened to shoot out along with whatever made semen make life.

Alas, preformationism lived long enough to become yet another misogynist villain. When Antonie van Leeuwenhoek first looked down into his microscope and saw microbes wriggling about in the 1670s, he also observed human sperm. And he quickly posited that these animalcules—which, rather adorably, is what he called all microbes—seemed lively enough to have tiny men inside them. He saw "all manner of great and small vessels, so various and so numerous that I misdoubt me not that they be nerves, arteries and veins . . . And when I saw them, I felt convinced that, in no full-grown body, are there any vessels which may not be found likewise in sound semen."

First of all, ew.

Leeuwenhoek would later backpedal his whole sperm-is-full-of-veins concept, but preformationism really cracked up from there. While ovists merely mistook human egg cells for literal eggs, spermists saw in semen a perfect delivery system for armies of minuscule menfolk.[3] Spermism supposed that a father essentially fired a tiny version of himself into the open vessel of mother, which, as I'm sure you'll agree, is just some *major goddamn bullshit*.

Darling preformationism's highest and lowest point came in tandem with the introduction of panspermism. If sperm contained tiny man babies—which only grew into women if the womb was defective and malformed, naturally—what happened to even less fortunate ejaculations? Was not every sperm sacred? What made a nasty lady womb so much better at fostering wee gentlemen than all the other places a man could ejaculate in or onto? In the words of feminist icon Elle Woods, "Why now? Why this sperm?"

Perhaps, some scholars supposed, errant ejaculations simply scattered onto the wind, taking hold and making life wherever a suitable host presented itself. Flowers, trees, birds, bees: anything alive might just stem from the nocturnal emissions of a human male. Many male scientists seem to have believed this wholeheartedly, inasmuch as it excused them from the ethical quandary of masturbating out tiny people with no hope of survival.

And yet, it is doubtful that any unwed pregnant woman ever successfully used the panspermism defense to avoid ostracization. Science, thou art a fickle beast. And also a sexist one.

Many theories die by a thousand paper cuts in the form of incremental findings and contradictory pieces of evidence. Preformationism adapted and survived for decades. It was not weak (like the sperm used to conceive women). But it dropped dead as a doornail in the face of cell theory in the mid-nineteenth century, which allowed scientists to finally conceive of a way for bits of people to grow up into whole-ass people.

Preformationism was not logical. But dammit, it was logical *enough*—at least until our microscopes weren't utter crap. It will always have a special place nested deep, deep in the gonads of every soul on earth.

A SERIES OF TUBES

It won't surprise you to hear that men got conception even more wrong before the dawn of microscopy. For example, in *The Generation of Animals*, written in Athens near the end of the fourth century BC, Aristotle details a fascinating test for female fertility. Simply stick a scented cotton cylinder into the vagina, wait for some reasonable length of time, and then check to see if she's started exhaling the odor of the diagnostic tampon through her mouth. Barring that, the physician might look for signs that her eyes or saliva had taken on some new hue as a result of the smelly insertion.

Aristotle believed that male and female "semen" had to come together to create a child (which is, in his defense, kind of right, except that he thought lady semen was menstrual blood). This test, then, would "demonstrate" whether a patient's body provided a clear pathway through which these semen could sail. If her vagina was not directly connected to her eyeballs by way of an internal breezeway, she was barren. One of my mother's patients used strikingly similar logic in the 1990s when she declared angrily that her newborn was covered in the Pepto-Bismol the nurses had given her during labor. She was actually seeing *Vernix caseosa* (Latin for cheesy varnish—yum), which is a waxy substance that coats a fetus's skin during the third trimester.

I can promise that whatever you eat and drink while pregnant, your baby will not come out covered in it. You are not an earthworm, and your internal tubing is not a straight shot from mouth to vagina.

THE WANDERING WOMB

You've probably heard this one before: the term "hysteria" originally referred to a whole general assortment of crazy lady symptoms, like having emotions and hating men, that supposedly stemmed from the uterus. You may even already know that, when these diagnoses first came about in ancient Greece, the issues were thought to result from the womb's migration throughout the body.

But I promise the story gets better.

According to Aretaeus of Cappadocia, a second-century Greek physician, the uterus essentially snuffled around inside you like some kind of rodent. In addition to making "altogether erratic" movements "hither and tither in the flanks," he noted that the organ "delights in fragrant smells, and advances towards them; and it has an aversion to fetid smells, and flees from them." On the whole, he claimed, the womb was like its own living creature.[4]

There are signs that this belief originated further back in Greek antiquity. Works by Plato circa 360 BC refer to uteri as *zōion*, or wild animals. He writes that these internal beasts become distressed and start to wander if left fruitless for too long.[5] Hippocrates, who was doing the whole father of medicine thing around the same time Plato was doing the whole father of Western philosophy thing, also noted that suffocation could arise if the womb drifted to and lingered by the diaphragm.

It was only around fifty years after Plato's death that dissections became a common way of studying human anatomy, which should have put the whole wandering ute theory to rest. After all, in a healthy body the uterus is decidedly not mobile in nature. Forgive me for not giving you an in-depth anatomy lesson. But, to make a long story short, the uterus, ovaries, and fallopian tubes are all blanketed in something called the broad ligament, which is a sort of sheet of peritoneum—the membrane that lines the abdominal cavity—that serves to connect them to the pelvic walls and floor. Several ligaments also help keep them steady. And the uterus itself is, of course, connected to the extremely stationary vagina by way of the strong, thick walls of the cervix. Like all your organs, the uterus is designed to hold up to some jiggling. That doesn't mean it goes wandering.

This would have been apparent to anyone who looked inside a uterus-having corpse, *and yet* . . . Centuries later, Soranus of Ephesus and Galen did argue against the notion that the uterus was a beast with a mind of its own—but they were outliers. The prevailing view was still that the womb could hit the road at any time. Indeed, Soranus spent most of his time figuring out a way to explain how the uterus might still shift enough to cause hysteria, despite his understanding that its ligaments would keep it from migrating through the whole body.

Depending on how mobile your ancient Greek physician thought your pesky uterus was, his prescription for your ennui could run the gamut from fumigation (to urge the organ back into position by way of tempting or repelling it with odor) to early marriage (as there was thought to be no womb more content than a pregnant one). Basically, the uterus was a free-spirited vagabond that might be persuaded to settle down with the love of a good man and a place to finally hang its hat.

The belief that women could only be healthy once impregnated was not universal. In the 400s in China, doctors noted that women who had children too young could put themselves in peril of lifelong sickness.[6] Chu Cheng, a physician of the fifth-century Southern Qi dynasty in China, went so far as to say that frequent childbirth and nursing at any age could "wither blood" and kill you. He advised both men and women to hold off marriage and reproduction for some years after they actually became capable of it. In his mind, women trying to conceive before age twenty and men marrying before age thirty would lead to a failure to achieve pregnancy at best and a frail infant and dead mother in the majority of cases.

Believe it or not, despite the evidence provided by a mounting pile of dissected corpses, the belief in wombs with wanderlust persisted until fairly recent history. In 1602, English chemist Edward Jorden testified as much at the trial of Elizabeth Jackson. This woman was accused of cursing her teenage neighbor Mary Glover with such symptoms as blindness, dumbness, convulsions, and swelling. Jackson was indeed convicted and imprisoned. But Jorden offered an impassioned defense, which he turned into a pamphlet the following year, printed in English when most medical texts were in Latin and inaccessible to the public.[7] In the pamphlet and on the stand, Jorden argued that Jackson and other "witches" are actually suffering from *passio hysterica*, the "suffocation of the mother," where "mother" refers to the womb. Without the "benefit of marriage," he wrote, a "congestion of humors" can send the womb wandering, and "perturbations of the minde" can cause physical symptoms like "frenzies," "convulsions," and "weeping."

By the time Sigmund Freud and Josef Breuer published *Studies on Hysteria* in 1895, most doctors thought the uterus-to-mental-illness connection was likely to be a bit less whimsical, if it existed at all. Some proposed that living in modern society caused stress or hindered a woman's ability to feed her natural need for femininity, which might have had impacts on both the reproductive organs and the mind. Freud and Breuer took things a step further by arguing that the symptoms associated with hysteria were actually the physical and psychological manifestation of sexual trauma experienced during childhood and had nothing to do with the health or movement of specific organs at all.[8] Freud would eventually pivot to saying the trouble lay in repressed childhood fantasies to explain all the folks who had mental health issues without having suffered profound trauma during youth, and then he'd say a lot of other stuff. But this isn't a book about Freud. Sorry! Still, around that time the direct connection between wombs and emotional turmoil was finally and decidedly severed in most medical circles.

Hysteria may be a particularly bonkers example, but history is full of plenty of other off-the-wall beliefs about pregnancy and fertility. Here are just a few of my favorites.

YOU HAD TO PUKE TO PROCREATE

The Kahun Gynaecological Papyrus is the earliest known gynaecological text, dating to around 1800 BC.[9] And it features several suggestions for evaluating fertility. Some of them are hard to puzzle out because the papyrus is literally in pieces and its translations are full of ellipses, question marks, and word fragments. For instance, we know that there were fertility tests involving onions, fresh oil, distended or limp female innards, the calf of Horus, a woman's lip, and a woman's nostrils, to name just a few examples, but the exact processes to evaluate these body parts or employ these ingredients are lost to time.

One of these procedures is just about clear enough to parse: Sit the woman on dirt smeared with dregs of sweet beer, put dates . . . somewhere (in her mouth? We hope?), and wait for her to "eject" from her mouth. Is she vomiting? Is she just spitting the dates out? Unclear. But every "ejection" promises one future birth, and a woman who expels nothing will be barren.

YOU HAD TO KEEP YOUR UTERUS FROM FALLING OUT

Doctors in the nineteenth century believed that vigorous sport could shake a woman's womb loose.[10] This would endanger her mental health, not to mention her future childbearing abilities. Nineteenth-century physicians also claimed that riding a bike would give women clenched jaws and bulging eyes, a condition they called "bicycle face."[11] It seems pretty clear their main concern was keeping girls from doing cool shit.

Now, uterine prolapse is a quite ordinary condition in which the uterus sags down into the vagina or even peeks out of it.[12] Ironically, rather than a sign of barrenness, this is most common after multiple pregnancies, especially when a person has lost muscle tone due to age. I can't promise that vigorous sport and family planning will keep your womb from wandering as you mature. But it's not going to hurt (especially if your workout routine includes a Kegel regimen).

FEMALE ORGASMS ARE KEY TO CONCEPTION

Born in 1098, Hildegard of Bingen—the lustful leprosy and poisoned sperm-blood lady—was a German nun, composer, theologian, and academic leagues ahead of her time. She also, somehow, seems to have had a great conceptual understanding of the female orgasm:[13]

> When a woman is making love with a man, a sense of heat in her brain, which brings with it sensual delight, communicates the taste of that delight during the act and summons forth the emission of man's seed. And when the seed has fallen into its place, that vehement heat descending from her brain draws the seed to itself and holds it, and soon the woman's sexual organs contract, and all the parts that are ready to open up during the time of menstruation now close, in the same way as a strong man can hold something enclosed in his fist.

Whether Hildegard got word of this whole orgasm thing from women on the street or she experienced it for her own dang self, we'll probably never know. But the medieval belief that a woman's, uh, brain heat is key to conception has persisted. In modern times, research on this question is generally framed around finding the *purpose* of a female orgasm. The utility of a male ejaculation is obvious: it's how sex turns into a baby-making activity. But sex seems decidedly unpleasant for the females of many species, and the fact that non-penis-having humans can *quite* enjoy the experience of copulation continues to blow many scientists' minds to this day.

While studies have come out with mixed results, the general consensus is that a female orgasm isn't particularly important to successful conception and definitely isn't necessary in every case.[14] Having an orgasm is *rarely* a bad idea, *especially* if you're having sex. But you should not stress out overly about making one happen (especially since this is a great way to guarantee you won't have any fun at all).

One current theory for why the female orgasm exists is that it's just an evolutionary holdover from the adaptations that gave us the male orgasm: like male nipples, but in reverse. Another theory is that orgasms have the benefit of making women more likely to have more sex and therefore more babies.

Sometimes folks take ancient notions about the female orgasm and really get them twisted. In 2012, former US representative Todd Akin infamously claimed that abortion shouldn't even be allowed in instances of rape. The female body, he said, had ways of trying to "shut that whole thing down," if the assault was "legitimate." In other words: you can only get pregnant if you enjoy it.

This is just about as wrong as a statement can be.

First of all, as already established, having an orgasm does not seem to affect conception; nor does any other metric of arousal. Second, while research on sexual assault is unfortunately scarce, some studies suggest that as many as 5 percent of rape survivors who come forward report having had an orgasm.

Some experts say that, based on anecdotal evidence from survivors they've treated, this figure seems low.

An orgasm can increase the shame, confusion, and anguish an assault survivor feels, but it has nothing to do with how "legitimate" their rape was.[15] For reasons we don't entirely understand—perhaps as a protective mechanism or perhaps due to some unfortunate quirk of biology—it is possible, and perhaps even common, for your body to physically react to stimulation that your mind finds repulsive or terrifying. Having an orgasm doesn't mean you weren't raped, and it shouldn't keep you from seeking help.

YOU CAN THINK YOUR WAY INTO
DELIVERING ANOTHER SPECIES

In 1726, a peasant woman named Mary Toft gave birth to a baby rabbit. It wasn't alive—it actually came out in a jumble of dismembered pieces, so I'm using the term "gave birth" loosely—but it was indeed a bunny, and it did indeed come out of her vagina. There also might have been a wee bit of tabby in there. And an eel that the tabby had eaten. These deliveries continued in the weeks to come.[16]

Scholars now suspect that Toft may have been manipulated into perpetrating this con by another party, but whatever her motivations, it was obviously a ruse.[17] Toft or an accomplice had been placing various mammalian bits and bobs up inside of her so that she could appear to deliver them out of her birth canal. It's possible Toft simply squeezed these animal chunks out of her vagina, though it's also been suggested that they were pushed into her cervix, which may have been softened during a recent miscarriage. I am not ashamed to say that my whole body clenches up in sympathetic pain and revulsion whenever I write that theory down. Anyway, wherever Toft actually stored those chunks, her process seems to have fooled both her local physician and several delegates sent by King George I, including the royal surgeon. Toft was only caught when, once moved to a new location far from home for observation, she was caught soliciting rabbit deliveries to her room.

But the most fascinating thing about this hippity-hoppity prank on the medical establishment is *why* it was thought to be plausible. Toft claimed that the trouble had started several months before her monstrous labor, when she'd been startled by a hare while working in the fields. She and another woman tried to catch it, she recalled, but they couldn't manage to chase it down—and then they spotted a *second* rabbit that also eluded them. So transfixed was she by the teasing critters that she dreamed of them that night. Toft claimed she had consequently lusted over rabbit meat for the duration of her pregnancy but had been unable to indulge due to her extreme poverty.

Remember how scientists used to think that sperm was full of tiny men who just shot into their mothers' wombs? Many of the same "experts" believed in something called "maternal impression" to help explain why people weren't born as copies of their dads—and to provide a tidy, mom-blaming explanation for any birth defects.[18] This belief dates back to ancient Greece, where stories were told about children being born particularly beautiful (or with an unexpected and inconvenient skin color) after their parents gazed upon statues or paintings of the gods. Some scientists were starting to side-eye the notion by Toft's day, but others upheld the idea for decades to come. Joseph Carey Merrick, more commonly known as the Elephant Man, lived a full century later but still blamed his distinctive growths on his mother's having been frightened by a circus animal.

There's a surprising kernel of roundabout truth in there. Researchers now know that various circumstances of our gestations can affect epigenetics, or the way certain genes are expressed. This is a pretty new area of research, but it seems clear, for example, that experiencing extreme stress during pregnancy can affect your descendants for generations to come.[19] But none of these potential effects are so obvious or magical as historically presumed. Feel free to look at as many elephants as you want.

IT ALL COMES DOWN TO THE LADIES

You may have noticed that most of the fertility tricks and tips outlined above have to do with people with uteruses. That's just kind of been the party line for most of history. In *The Generation of Animals* (which is, disappointingly, not a what's-up-with-millennials-style takedown of the youth but rather a book about how animals are generated) Aristotle himself notes that if a couple can't conceive, the explanation may lie within the man, the woman, or both parties.[20] But after stating this, he fails to provide any suggestions for how to treat male infertility. Instead, he focuses on how doctors might investigate the female body. Ancient Greek texts may offer men some tips and tricks for *helping* their partner conceive: when and how to have sex with them, for instance, and which testicle they should tie up if they want to have a son (the left one). But infertility is presented as a problem most easily fixed by treating the female party.[21]

It's not as if no one knew men might sometimes be incapable of having children. In *The Hidden Treasures of the Art of Physick* (1659), John Tanner advised doctors not to assume women were to blame. "It is known thus," he wrote, "if the man be unable to raise his yard, if he want Sperm, if he hath a swelling in his Stones, or if he have the Running of the Reins, he is not fit for Venus School." Tanner goes on to add that if a man is particularly effeminate, lacks a beard, takes a long time to come, doesn't seem to enjoy sex, or ejaculates cold semen (blech!), he's probably shooting blanks.

But even in the 1700s, men were in the habit of blaming their wives for any procreative difficulties . . . often despite all evidence to the contrary. Take George Washington: When he and Martha married at the shared age of twenty-seven, she was a widow who'd borne four children in just eight years. The pair of them never had any additional kids together. But a letter he wrote at the age of fifty-four suggests our Founding Father blamed that fact on his demonstrably fertile wife: "If Mrs. Washington should survive me there is a moral certainty of my dying without issue and should I be longest liver, the matter in my opinion is almost as certain; for whilst I retain the reasoning faculties, I shall never marry a girl; and it is not probable that I should have children by a woman of an age suitable to my own, should I be disposed to enter in a second marriage."[22]

George gets credit here for not chomping at the bit to go marry a girl young enough to be his daughter, I guess. But his assumption that he'd be able to knock up a nubile spouse—when his wife was, by all accounts, quite fecund when they wed—is perplexing to say the least.

When microscopes finally got good enough in the mid-nineteenth century for us to really grok how sperm and egg came together to make conception happen, there was finally a target for the concerns about male fertility that had so long been left on the back burner. In 1866, Paolo Mantegazza made the first connections between the characteristics of a person's sperm and their ability to impregnate someone.[23] He noted that higher semen volume tended to mean more baby-making power and did some experiments on the role of temperature in seminal success too. Mantegazza came up with the idea of a sperm bank, noting that it might be a way for men heading off to war to help ensure their wives could conceive heirs even if they died (though it would take decades more for scientists to figure out how to effectively freeze human semen to keep it viable).[24]

By the twentieth century, it was widely accepted that sperm could vary widely enough in quality to keep a couple from conceiving. Today, any fertility specialist worth their salt will investigate each half of a couple equally and offer interventions to either or both members of the baby-making equation as needed. Thank God we're not still shoving dates into our mouths and hoping we'll be lucky enough to puke.

Chapter Eight

HAVE WE ALWAYS USED BIRTH CONTROL?

. . . .

In which crocodile dung goes into some unspeakable places, Casanova loses all of your respect, and people sneeze semen everywhere.

. . . .

There is every indication that abortion is an absolutely universal phenomenon, and that it is impossible even to construct an imaginary social system in which no woman would ever feel at least impelled to abort.

—George Devereux, "A Study of Abortion in Primitive Societies: A Typological, Distributional, and Dynamic Analysis of the Prevention of Birth in 400 Preindustrial Societies," 1955.

CONTRARY TO WHAT PERIODIC PEAKS OF MORAL OUTRAGE MIGHT HAVE you believe, baby making has always been pretty low on our sexual to-do list. Don't believe me? Just ask the women from some four thousand years ago who spent their days shoving crocodile dung up into their vaginas.

As discussed in our chapter on reproduction, humans were blissfully unaware of the exact mechanics of human fertilization until the late 1800s and didn't even spot sperm for the first time until 1677. But long before we'd begun to investigate the little swimmers that help us wriggle our way into existence, people got wise (relatively speaking, anyway) to the fact that when two become one, they could then become three.

We don't have any way of knowing when and how this became common knowledge. That's because the idea that sex and babies are linked is *probably* older than our own species. A hominid wouldn't need that much cognitive prowess to put two and two together. And communities were probably more successful when they told their kids that flagrant boning might leave them with another mouth to feed. In fact, it's likely that the first widely practiced method of birth control was simply breastfeeding; in a brilliant evolutionary bid to make sure mamas have enough resources to rear their young to maturity, regular nursing triggers hormones about as effective as modern contraceptive pills, at least for the first six months.

But for people who weren't interested in having even one kid (or nursing them until they went off to college), the first evidence of effective birth control crops up in texts from ancient Egypt.

Sometime between the realization that sex made babies and the earliest days of recorded history, folks decided semen was the thing to avoid. This was smart: if ejaculate can't make its way into the cervix, pregnancy can't happen. So Egyptian women in 1850 BC combined bits of honey, salts, and sometimes crocodile dung to create spermicidal plugs. The acidity of the poop and other ingredients, combined with the stickiness of the honey and the physical barrier of the plug itself, would have been fairly effective at blocking and killing sperm. Meanwhile, although we're not sure if they knew it at the time, the powerful antimicrobial properties of honey would have helped keep the gooey suppository from causing a dangerous infection. By the 1500s BC, ground-up acacia leaves (which we now know produce lactic acid, which inhibits sperm motility) were a crucial component.

Archaeologists have even found recipes for abortifacient concoctions inside ancient tombs, perhaps indicating that Egyptians counted on having a pretty freaky afterlife and didn't want any undead kids to put a damper on it.

You may not quite buy that early humans knew how to harness breastfeeding as a contraceptive. Even so, with the records of these (theoretically) effective Egyptian contraceptives, that makes almost four thousand years of *documented* attempts to do the crime without doing the time. And we haven't stopped since. In 1997, psychologist Craig Hill set out to determine the most common motivating factors behind human sexual intercourse. He presented eight typical motivations; getting pregnant got ranked dead last.[1]

THE LACEDAEMONIAN LEAP

The ancient Greeks believed that semen somehow literally *turned into* a human, which surely caused even more Sturm und Drang over the question of when life began than we grapple with today (the answer, according to the modern medical establishment, is two weeks before the parent actually gets pregnant, as we learned recently). Despite this, historical records show such robust references to tips and tricks for ending unwanted pregnancies that it seems unlikely abortion was much stigmatized in ancient Greece and Rome.

That's not to say there were no naysayers; according to the Greek physician Soranus of Ephesus, the famous Hippocrates was antiabortion. It's worth noting, however, that Soranus lived several centuries after Hippocrates died—and that there are records of Hippocrates helping to "dislodge" the "seed" of a days-old pregnancy from at least one young patient.

We may not know Hippocrates's motivation in helping this one girl get unpregnant (though a text attributed to him suggests it was to keep the "singing girl" from losing her financial value). But we know the method he claimed to use . . . and it sounds like a teen dance craze: the Lacedaemonian leap.

You put your hands on your hips and bring your knees in tight—but in lieu of pelvic thrusts (after all, that's what got you here in the first place), you jump into the air. For some reason it is important that the feet touch the buttocks while the jumper is airborne. A few forceful jumps and butt kicks, Hippocrates claimed, would handily dislodge a growing embryo in mere moments.

There are a few things to unpack here, friends. First of all, while butt kicks are wonderful cardio and can really help you jack up those quads, no amount of exercise is particularly likely to trigger a miscarriage. Second, I feel it necessary to inform you that, according to the historical text recounting this "medical" procedure, Hippocrates claimed the "seed" had dropped from his young, leaping patient with an audible plop. Embryos are certainly not up to plopping weight in the first few weeks of gestation. Given the general state of confusion about when and how the stuff inside ejaculate actually became a baby, it's not much of a stretch to imagine that what Hippocrates really heard expelled from his patient was a whole lot of spunk.

We now know that healthy semen is full of eager little sperm cells numbering in the millions. And, as such, we know that no amount of postcoital cleansing or cervical spelunking can guarantee none of these genetic messengers will reach their target. But as previously mentioned in our pregnancy chapter, until we spotted sperm under a microscope in the 1600s, many assumed that jizz itself helped create a fetus by literally congealing into shape. Given that belief, it's easy to understand why ancient Greek and Roman physicians recommended that women avoid pregnancy by simply shooting the stuff right out of their vaginas—either by jumping up and down or by sneezing vigorously.

Soranus, a Greek physician who worked in Rome in the first and second centuries AD, suggested the following: "At the critical moment of coitus when the man is about to discharge the seed, the woman must hold her breath and draw herself away a little, so that the seed may not be hurled too deep into the cavity of the uterus. And getting up immediately and squatting down, she should induce sneezing and carefully wipe the vagina all round; she might even drink something cold."

Let me be clear: This did not work. No amount of sneezing will keep you from getting pregnant unless your snot rockets manage to deter all potential sexual partners in the first place.

In the same text, Soranus suggested that women also cover their cervixes before sex using pastes made of various resins, oils, and wool fibers. This probably worked a lot better, for the same reason that ancient Egyptian women had so much success with gummy insertions made from sticky honey and acidic acacia leaves.

But jumping and sneezing must have, at the very least, made for a dramatic exit from any particularly regrettable one-night stands. And perhaps it kept participants too distracted (and tired) to go for a second round.

WHEN LIFE GIVES YOU LEMONS

Cervical caps pop up again and again in the history of sex—including in the escapades of Casanova. You might assume that being one of Giacomo Casanova's lovers was a glamorous affair. There's a reason his name has become synonymous with seduction and romance: his memoirs recount a level of wooing and regard for sexual satisfaction that still sounds shocking to most of us today. But let's clear up a few things.

For starters, while we have a tendency to assume all our ancestors were prudes, the eighteenth century was an *extremely horny time*, and Casanova was *not special for being horny*. There's also the fact that he gleefully admitted to seducing his teenage daughter, grooming and sleeping with at least one child, and buying a young orphan to keep as a sex slave. The word you're looking for, friends, is "yikes." More like Casa-no-thank-you.

Casanova also gets a lot of credit for popularizing or even inventing several methods of birth control. But there is nothing admirable about a man convincing his sexual partners to shove lemons up their hoo-has, which is how many claim Casanova came up with the idea of a cervical cap. Yes, blocking sperm from entering the cervix is a great way to prevent pregnancy, and yes, the acidity of lemon juice probably helped kill some of the swimmers that got close. But if the thought of lemon juice being squished up into your cervix during sex doesn't make your whole body pucker, I'll tell you what: I bet you don't have a cervix.

(Side note: Some historians say that women enslaved on American plantations used such hollowed-out citrus fruits as acidic contraceptive barriers as well.[2] But *they* were working with what they had, living under the constant threat of violence and assault and turning lemons into contraceptive lemonade, and Casanova was a rich white man during a time when it was great to be one in Europe. As such, I maintain that he was a jerk for imposing squirts of lemon juice on his partners' genitals.)

In reality, the rudimentary cervical cap Casanova claimed to have "invented" was no more innovative than gooey plugs concocted by ancient Egyptian women. And even those barriers weren't one-of-a-kind. The Egyptians' ancient contemporaries in India used clarified butter or thick oils as a smeared barrier over the opening of the cervix. The Talmud, meanwhile, recommended a spongy material called *mokh* soaked in vinegar for women too young or frail to get pregnant safely; this actually persisted as a common birth control method among Jewish women through the sixteenth century. Time and space are crammed with tales of women cramming things up against the cervix to prevent pregnancy: seaweed on Easter Island, bronze pessaries in ancient Rome, oiled paper discs for Japanese prostitutes throughout antiquity, carved wooden caps in America at the turn of the twentieth century, and, finally (thankfully), the silicone diaphragms smeared with spermicide enjoying a resurgence today.

And you can stop thanking that jabroni Casanova for male condoms, while you're at it. Thank the Edo-period Japanese men who capped their penises with little cages carved from tortoise shells, because *what the flying fuck*. Anyone can pick up an animal intestine and slip it on before sex; that's the preplastic equivalent of seeing a sandwich baggie and figuring it could probably get the job done. But affixing a hard shell to the top of your dick? Now *that's* a stroke of horrible, terrible genius.

The first mention of a condom in the historical record, in fact, comes from 3,000 BC; that means condoms show up more than a thousand years before the Egyptian dung plugs, albeit in a more whimsical fashion that has little to do with contraceptive efforts. They first appear in Crete, when the famous King Minos supposedly sought to protect his wife from the "serpents and scorpions" that emerged from his semen during sex.[3] The pests had apparently killed all of his favorite concubines, and celibacy wasn't an option for a man in need of an heir and several spares. The donning of a goat bladder did the trick, protecting his wife Pasiphaë from intravaginal scorpion infestations while allowing her to conceive eight children.

(If you're wondering whether the creepy crawlers were a masked reference to some kind of sexually transmitted infection, well, yeah, duh. But that idea actually raises some fascinating questions, because the microbes that cause STIs are *smaller* than sperm cells. Animal membranes like the ones used in modern lambskin condoms can keep baby makers at bay but don't reliably prevent the spread of disease. Pasiphaë's ability to conceive without being struck down by her husband's cursed ejaculate is only plausible if, first, that goat bladder had some little holes poked through it and, second, the bugs in his semen were *actual bugs*. Then again, some legends have it that Pasiphaë actually cursed Minos's semen herself to keep him from having sex with *other* women, so maybe the goat bladder condom was just to give her plausible deniability for somehow surviving their marital encounters. Alternatively, given that Pasiphaë was said to have cuckolded Minos with a literal bull to conceive the legendary Minotaur monster, maybe she simply avoided her husband's semen altogether and sought children by other means. Also, this story is made up, so let's not sweat over it too much.)

Men in ancient Egypt donned little linen sheaths, which sounds pretty futile until you realize they did it to prevent the spread of parasitic *worms*, not bacteria (or babies). What was it with ancient semen and actual bugs? We really do have it good today, folks. Ancient Romans upgraded to less-permeable animal skins and bladders, though they were primarily concerned with stopping the spread of syphilis; birth control would have been an unintentional side effect. In addition to the aforementioned tortoise-shell dick toppers in Japan, other Eastern nations used silk to fashion jaunty little penis coverings.

Back to dunking on Casanova: Condoms made great strides starting in the fifteenth century, as syphilis was raging hard and whores were whoring even harder. By the time Casanova hit the scene, their use was already ubiquitous. In fact, his memoirs suggest that as a young man he found the devices repulsive and refused to wear them. Eventually he decided that neurosyphilis was worse, which may have been the one thing he was ever right about.

In 1855, thanks to Charles Goodyear's innovations in rubber vulcanization—the process of mixing natural rubber and sulfur to form a stronger, more elastic material—we could finally cast aside animal guts in favor of a more effective material. Unfortunately, that material was about as thick as an inner tube.[4] Even more unfortunately, people had them custom-made and washed and reused them. Sexual health is great and all, but something about a reusable and rubbery rubber is nearly gross enough to make one miss the days of unchecked neurosyphilis.

The 1920 discovery of the rubber-derivative latex made for a much happier time, as the material is thinner and stretchier and doesn't make your penis look like it's wearing a tiny gimp suit. Trust me: we have nothing to complain about.

WHAT A DOUCHE!

Our attempts to sneeze our way to sterility may be well behind us, at least when it comes to advice your doctor will give you. But we have much more recent history with efforts to erase evidence of sex after the fact by simply cleaning up the mess. Modern women were smarter than their sneezing ancient counterparts, so they shot Lysol up their vaginas.

The year 1843 was an excellent time to be a douche. French obstetrician Maurice Éguisier designed a spring-loaded device that would automatically spray water through its hose, and women went wild for this hands-free sperm swirler. It looked like a toilet bowl brush holder attached to a garden hose and was about as good as those objects are at preventing pregnancy.

Freed from the tyranny of irksome bulbs and syringes, the next frontier in douche innovation focused on the liquid surprise inside. If water was good (it wasn't!), then wouldn't spermicidal chemicals be even better (they weren't!)? By the 1920s, doctors were bemoaning the ineffectiveness of douches as a birth control method and reporting negative side effects in patients who used them. But the douche was out of the nozzle: because women could buy such products in catalogues and department stores—even as antiobscenity laws forbade the spread of information about contraception—they were by far the most popular method of birth control in the United States.[5] And because they were a popular product, companies put a lot of effort into marketing them. Lysol was no exception.

Just imagine, it's 1950, and the first hormonal birth control pill is a few years from clearing clinical trials. You're a coiffed housewife hoping to avoid pregnancy. Condoms and withdrawal are fine, but you've been burned before—methods that rely on your husband's cooperation are less than ideal. Thank God for Lysol! It's advertised in your favorite magazines as promoting "marriage hygiene," which everyone knows means it'll keep your uterus spick-and-span and inhospitable to human life. So, just like your mother did before you, you follow all sexual encounters with a quick vaginal blast of the same astringent cleaner you use to make your toilet sparkle.

You weren't alone, and you're still not. While medical outcry about burns, scarring, and lasting cervical damage in the 1980s led Lysol to stop tacitly endorsing the medical use of its cleaning products, millions of women around the world still use a host of largely unregulated douches. Now that we know no amount of astringent rinsing can actually prevent pregnancy—of the millions of sperm cells present in each burst of ejaculate, at least a handful will make it safely into the cervix in mere seconds—the "hygienic" aspect of douching agents has gone from a wink-wink-nudge-nudge code to a serious marketing angle. Companies make customers believe that the smells and secretions of the vaginal area are disgusting symptoms in need of treatment and that staying fresh and clean with vaginal douches can provide a cure (and even, some products imply, keep you safe from disease).[6]

Let's get something straight: While the vaginal microbiome is complex and occasionally prone to acts of minor rebellion, the associated reproductive system is a self-cleaning oven. It's supposed to secrete things. It's supposed to have smells. Only when things seem out of the ordinary or a person is in pain or discomfort is medical intervention necessary (and, at least in my own experience, you almost always just need to stop sitting around in dirty yoga tights all day).

Ironically, the act of douching can actually help convince you that you *need* to douche.[7] Numerous studies have shown that rates of bacterial vaginosis (also known as stanky discharge) go up with regular douching, since rinsing these delicate tissues with anything but plain water can disrupt the area's pH or even kill off good microbes and allow bad ones to thrive. It may even increase the risk of sexually transmitted infection by way of leaving vaginal skin irritated and more prone to tear during sex, providing an easy point of entry for foreign invaders. In other words, regular rinsing can give you the kind of smells and alarm bells you've been taught to fear from your vagina, which can, in turn, cement your conviction that you need a special wash to keep your body in check.

Okay, okay, douching bad, natural vaginal flora good. But let's get back to your Lysol-loving 1950s persona. Did her DIY emergency contraception actually *work*?

There are indeed concoctions that can kill sperm, and something as toxic as Lysol is almost certainly among them. In 1985, a group of researchers at Harvard Medical School (inspired by one student's story about girls at her Catholic boarding school using various liquid spritzes to avoid pregnancy) tested the spermicidal properties of Coca-Cola.[8] They found that the beverage did a halfway decent job of crippling sperm within just a few minutes. Indeed, those early suppositories of crocodile dung and acacia leaves worked by killing off little swimmers, and those components were far less obviously harsh than a soda pop.

But the key to the efficacy of those ancient spermicidal systems wasn't just their toxicity; the best methods combined the antimicrobial agents with some physical barrier, like honey and cotton or claylike animal feces, to keep semen from reaching the cervix in the first place. Shooting a spermicide into the vaginal canal after the carnal act will miss the fastest sperm, as they'll already be safely sequestered inside the cervix. Am I saying that shoving honeyed crocodile scat into your vagina is probably less harmful than spraying your cervix with Lysol? Yes. But please don't do that either.

BOIL IT, MASH IT, PUT IT IN A STEW

In the beginning, there was a plant. And the plant was really good at keeping you from getting knocked up. But it went the way of the dodo, the passenger pigeon, and any hope of my grandchildren living on a planet that's not plagued by catastrophic climate change. Humans were so voracious as to completely destroy all traces of silphium on the planet.[9] That's how the story goes, anyhow.

Described as some sort of giant fennel by Pliny the Elder in AD 77, silphium was a hot-ticket item in all spheres of Greek and Roman life. Boil it, mash it, put it in a stew, use it as a compress for an infected dog bite, trick snakes into drinking wine laced with it until they exploded—what couldn't you do with one of these puppies? Julius Caesar was said to have found nearly a literal ton of the stuff in the public treasury chest in Rome during the civil war, and at least one city became fabulously wealthy from growing and selling sprouts.[10] But by the time Pliny was singing the praises of this particularly versatile vegetable, it had all but disappeared: he claimed, in his writing, that only a single stalk had ever been seen in his entire lifetime.

One of silphium's more intriguing purported properties was an ability to prevent pregnancy and bring on menstruation. Conveniently, it was also highly regarded as an aphrodisiac. Nothing says no-strings-attached sexy shenanigans like a big ol' hunka fennel, am I right?

Of course, it's impossible to know how silphium ranked in efficacy on a scale from sneezing out semen to surgical sterilization. As far as anyone can tell, we managed to mow it off the face of the earth. Many have been tempted to suggest this botanical disaster was the direct result of desperate, horny men and women hoarding the plant to support their bedroom habits. But the truth is that silphium was popular, hard to cultivate, and only grew wild in areas increasingly inhabited by sheep farmers (and their sheep). Consuming silphium clearly did not have a contraceptive effect on livestock, which happily grazed on the stuff until it was in dangerously short supply.

But do not be deceived into believing that all ancient methods of botanical contraception were foolish. It's fair enough to make fun of sneezing out semen and to glance askance at myths about Romans absolutely mowing through the entire world's supply of natural Plan B. But countless cultures throughout time and space relied on oral contraception before scientists ever thought to craft a birth control pill, and we have reason to believe that many of those supplements did what they were supposed to do. This is hard to state with certainty: while some scholars point to particular dips or plateaus in birth rates at certain points in history as evidence that everyday people were more in control of their fertility than generally assumed, others have pointed out major flaws in such arguments.[11] I'm going to go out on a limb and wager that the truth is pretty mundane: a lot of herbal contraceptives probably failed to work or at least didn't work that well, but some of them likely did—and we have the chemical evidence to back them up.

In 1941, the US Department of Agriculture published a record of medicinal plant uses among Indigenous peoples in Nevada. The record notes that one settlement of the Shoshone tribe believed a plant called *Lithospermum*, or stoneseed, had contraceptive properties when taken regularly mixed in water. According to a study published in *Endocrinology* a decade later, the academic world's sudden awareness of this Indigenous knowledge led to "extensive experimental work" in the 1940s and 1950s. In one study on mice from around that time, the plant *Lithospermum ruderale* did indeed manage to halt the menstrual cycle and induce temporary infertility.[12] Either fortunately or unfortunately, depending on your aims, research suggests this effect can become permanent if the plant is taken regularly for an extended period. So it's not shocking that scientists didn't rush to put it into pill form. Even so, it is surprising and disappointing that the research boom seems to have died down almost as soon as it started.

Because so little work has been done on the potential downsides of taking stoneseed or how best to administer it to prevent pregnancy, trying it for yourself is inadvisable unless you have access to the oral tradition of people who have honed its use over generations. In any case, the Shoshone probably weren't exclusive keepers of the knowledge of this plant's power. A 1966 survey of edible and medicinal plants used by tribes living on Montana's Flathead Indian Reservation notes that *Lithospermum* was taken for diarrhea.[13] It's hard to believe that communities could regularly ingest this plant without noticing (and perhaps even utilizing) its antifertility side effects.

Another contraceptive that seems to have stood the test of time is Queen Anne's lace, or wild carrot. According to a 2014 review in the *Australian Journal of Herbal Medicine*, the seeds of *Daucus carota* have been used to control fertility for some two thousand years.[14] A cadre of ancient Greek writers cited the seed as an emmenagogue or abortifacient. *Culpeper's Complete Herbal*, the go-to medical resource for many common folk when published in 1653, suggested these carrot seeds to both promote and hinder fertility by regulating the menstrual cycle. The plant has also been used historically to induce abortions, prevent pregnancy, and treat uterine pain throughout India, where it's known as *gajar*.

The same 2014 review highlighted several potential mechanisms by which Queen Anne's lace might prevent pregnancy, all gleaned from studies in animals. But the study authors also found heaps of anecdotal evidence and even informal studies on its use in humans. Still, they concluded that more research was needed to understand just how well it works and how safe it is for people to take in the long term.

A 2004 overview of folkloric contraceptives found that many of the small percentage that have been studied appeared useless in the lab.[15] Even so, the paper's authors found that several dozen have shown promise or have even had useful active compounds isolated for further study. Indeed, botanicals could hold the key to some of the high-tech birth control options of our egalitarian sci-fi dreams. In 2017, researchers found that compounds found in plants such as mango, dandelion root, aloe vera, and *Tripterygium wilfordii*, or "thunder duke vine," could act as a sort of molecular condom.[16] Pristimerin and lupeol were both found to hinder sperm motility by blocking the hormone progesterone.[17] More trials are currently underway, but these compounds could potentially prevent pregnancy when taken by either the ovulating or the ejaculating party. If the sperm-producing partner took them before sex, their swimmers would come out of the gate with laggy, ineffective tails. Taking a medication made with these compounds would enable the body of the ovulating half of the pair to cripple the incoming sperm by limiting their access to progesterone during the last leg of their trip.

More research is needed before we can rely on fruits and veg for all our contraceptive needs. But there's reason to believe the natural world is brimming with chemicals that could help control fertility in the future.

Time for a wee disclaimer, lads and lasses and duckies. Personally, I think it's kind of bizarre that herbal options get so little attention from the medical establishment. *Even so*, the information in this book is *not a sufficient starter guide* by any means. If you decide to proceed in this manner, you should take care. You'll need to consult a seasoned herbalist or a guide written by one, do research to make sure you don't have any contraindications in terms of underlying conditions or prescription medication, and source trustworthy supplements or learn to make your own.

Herbal contraceptives deserve a better seat at the table than they've gotten in the modern day. Still, medical regulators' lack of attention to them means you take them at your own risk. Pennyroyal, for instance, is widely cited in historical and folk medicine texts as an effective birth control or abortifacient, but it can absolutely be toxic if you take it in the wrong form, take too much of it, or take it along with other substances that tax your liver. You shouldn't take herbs carelessly just because they're "natural." (You know what else is natural? Dying.) Nor should you assume a high level of efficacy.

If you need an abortion and are able to access one in a medical setting, please make every effort to do so. A medical clinic will be able to determine how far along you are and ensure that your termination has been successful and is complete before sending you on your way. Both of these things are crucial to ensuring your health and safety.

AN INTRAUTERINE *WHAT NOW?*

Today, intrauterine devices (IUDs) are some of the most effective birth control options out there. This is largely because of the lack of room for user error; once they're implanted, there's not much you can do to *use* them wrong (though they can fall out, which obviously takes their efficacy down to zero pretty quickly). Modern IUDs work either by passively releasing hormones right where they're needed to halt ovulation and thicken the mucus on the cervix, which minimizes the risk of conception, or by killing sperm by exposing it to copper. IUDs aren't perfect for everyone: the insertion itself can be incredibly painful, copper devices can increase menstrual flow and cramping, and hormonal options can raise the risk of potentially dangerous cysts in some users. Still, on the whole, IUDs are kind of miraculous and a great way to ensure you'll almost certainly not get pregnant for anywhere from three to twelve years, depending on the type you choose. But IUDs weren't always so magical.

The woozy-making history of intrauterine devices begins in the late 1800s, when doctors asked a terrible question: What if pessaries, those precursors to diaphragms that blocked sperm by physically capping the cervix, also went *into* the uterus?

Unlike modern IUDs, which are squished through a dilated cervix but then exist entirely inside the womb (save for a nifty little string to aid with removal), the gold stemmed pessaries developed in Germany in the 1880s kept the opening of the cervix permanently occupied.[18] The devices would probably have helped decrease fertility drastically, both by blocking sperm and by inducing an inflammatory response in the uterus that would have made it a hostile place for a zygote. But any device that crosses from the vagina into the uterus leaves the body prone to all sorts of infections; your outer bits are frequently exposed to microbes that your inner bits are woefully unprepared for. Plus, just, ouch? As someone who's been on the receiving end of three IUDs over the span of the last decade (don't judge me, I'm indecisive), the thought of having a piece of metal yanking the ol' ute around for a prolonged period is enough to make me want to put my head between my knees.

While stem pessary designs varied, most of them were thick-necked enough to make one say, "Oh no."[19] The 1900s saw the advent of several devices that used catgut string or silkworm strands to form the "innie" portion of these contraceptives, which probably made them gentler on the cervix. But the continued presence of an "outie" in the vagina made them extremely risky for anyone who might catch an STI, which could wiggle its way up to the rest of the reproductive system. Only in the 1920s did designs meant to stay completely inside the uterus come about.

In the 1960s, with the help of a more widespread acceptance of contraception and the availability of plastics to form more flexible devices, the concept of the IUD really took off. In the 1970s, nearly 10 percent of women who used contraception in the United States used an IUD. But in 2002, that rate had dropped to 1 percent.[20] Why? Because the IUD's history isn't a straight line to success; it's a jagged, stingray-shaped roller coaster of thrilling highs and deadly lows.

In the 1970s, the most popular IUD on the US market was the crabby Dalkon Shield. (Literally crabby—it looked like a pinchy crustacean.) Doctors prescribed it to more than 2.2 million patients in its first three years of sale alone. Even during this period, however, there were signs of trouble: people were reporting serious infections and miscarriages associated with the device. While the shield may have protected some users from unwanted offspring, others faced dangerous pregnancies that caught them completely unawares. A 1974 study by the Centers for Disease

Control and Prevention—which followed nearly seventeen thousand medical practitioners to get data on health outcomes related to the Dalkon Shield and other IUDs over the course of six months—found the device increased the risk of bacterial infection during pregnancy. A follow-up in 1975 concluded that Dalkon's device carried a higher risk of miscarriage-related death than other available IUDs.

That got the shield pulled from the market, but Dalkon didn't recall the millions of devices that had already been implanted. By the time the company filed for bankruptcy in 1985, it was facing lawsuits in every state and had paid out hundreds of millions in settlements to people claiming pelvic inflammatory disease, miscarriage, and infertility due to their product.

The oft-cited culprit behind the Dalkon Shield's deadly side effects was the woven string it left dangling in the vagina; some experts say the material's tendency to fray made it like a straw that sucked harmful bacteria up into the uterus.[21] Modern IUDs still come with strings attached to make it easy to tell when one has fallen out, as well as to ease removal. But today's devices are reeled back out using fishing-line-like filaments to minimize the risk of bacterial migration. Other researchers suspect that many of the shield's poor outcomes could have been due more to faulty insertion and placement than to its design.

Whatever the cause, all IUDs took a hit.[22] Even in 2018, with decades to regain public trust, surveys showed that just 13 percent of women using contraceptives relied on IUDs—the same percentage who used condoms, which are far less effective.[23]

Picking a birth control method is a complicated and personal decision, and an IUD might not be right for you. Modern IUDs like Mirena also regularly face lawsuits for rare but dangerous side effects.[24] No pharmaceutical product is entirely without risk, and IUDs are no exception. But if your doctor hasn't even given you the option, it's worth asking them why. They may be operating under outdated prejudices against the devices—or simply assuming that you are.

JUST SAY NO

Every method of birth control has two efficacy rates. First, you have a percentage for perfect use. If one hundred couples use condoms every time they have sex and use them *properly*, at the end of the year only around two of those couples should have conceived. But then there's a pesky figure called "typical use." Once you factor in all the room for human error—using expired condoms, keeping them where your cat can claw at them, carrying them around in your back pocket even though body heat degrades latex, putting them on upside down, not actually *using* them, as if using them *sometimes* actually counts for anything, you dingus—a whopping fifteen to eighteen of those one hundred couples will have spawned something or other by the time the year is out.

But there's no wider gap between perfect and typical use than when it comes to abstinence. It really, really works until the second it doesn't. But that sure hasn't stopped us from trying. Of all the antiquated methods of birth control that exist for us to roll our eyes at, promising to keep one's legs closed has had the most sticking power. In 2016, the US budget for "abstinence only until marriage" sex education in schools—which has been shown time and time again to be ineffective in combating both teen pregnancy and sexually transmitted infections—was increased to

$85 million per year.[25] As of 2014, the average high school health course spent less than eight hours talking about all STIs and pregnancy prevention combined, and 87 percent allowed guardians to exclude their children from even this coursework.

Studies suggest that folks on abstinence pledges have pretty much fifty-fifty odds of ending up pregnant in a year.[26] Flipping a coin to decide whether to have sex every time you're thinking about it sounds like a sexier option.

CONTRACEPTION'S DARK SECRET

Unfortunately, we can't talk about birth control without talking about Nazis. I'm sorry! Really. But if we don't circle back to Nazis, I'll be leaving you with a woefully incomplete sexual education. I wish I could promise these will be the only Nazis in this book, but I can't. The history of Western medicine is pretty littered with them.

I am going to ask you to hold two truths in your heart simultaneously. Here's truth number one: access to birth control should be considered a basic human right. It's the only way to ensure an individual can have autonomy over their body and the freedom to make decisions about what kind of life they want to live. For many people, it is a matter of life and death.

Here is truth number two: we probably wouldn't have the modern contraception options available to us today if not for some incredibly dubiously motivated work by some unabashedly classist, racist people.

In 2020, Planned Parenthood of Greater New York announced that it would remove the name of founder Margaret Sanger from its Lower Manhattan clinic.[27] Sanger is largely and rightfully credited with making access to birth control a mainstream desire and reality. But she also dabbled in eugenics, which is the idea that human reproduction can and should be controlled to yield superior offspring.

Sanger was by no means alone. The American eugenics movement was considered mainstream science in the early twentieth century. Biologists like Charles Davenport were actively trying to convince the country that they could influence human nature by keeping certain people from breeding; psychologists like Henry H. Goddard were putting together fabulist case studies on degenerate families full of the "feeble-minded" as evidence that some lineages just didn't deserve to persist; doctors like Harry Clay Sharp were setting laws into motion that would lead to the involuntary sterilization of tens of thousands of people in the name of the public good. Everyday citizens were attending state fairs to compete in "better baby competitions," where children were weighed and measured like livestock to show the superiority of their bloodlines. Researchers published papers, news articles, and even ads imploring members of the public not to weaken the race by offering support to people with disabilities. Even premature babies were considered a risk to genetic integrity; the first American neonatal intensive care units were built on ocean boardwalks and paid for with the ticket fees of curious visitors, because hospitals couldn't be bothered to spend resources keeping weaklings in the gene pool.[28] None of this bears scientific scrutiny, by the way, and the assumptions propped up by eugenicist scholars largely boil down to the belief that some ethnic groups are superior to others and that poverty, criminality, and illiteracy are the results of your parentage (not true) as opposed to natural consequences of marginalization and oppression.

However Sanger may have felt about eugenics, which is a matter of some debate, she used the popularity of the movement to get birth control options into the hands of American women.[29] She and other early proponents sang the praises of "social hygiene": a vision of families small enough to stay healthy, tidy, and well behaved under the crush of US capitalism. In mainstreaming birth control, she also helped normalize the idea that some people should have more children than others and that certain people could only breed degeneracy into the population.

Believe it or not, Adolf Hitler et al. read American books on eugenics for inspiration.[30] During the course of World War II the American medical establishment started to feel a little *iffy* about its association with Nazism, and eugenics became more of a subtext.

Sanger's work did have a payoff. Along with physician John Rock, she helped pioneer the first clinical trial of an oral contraceptive pill in 1954.[31] But the path of medical research never did run smooth; Rock ended up evading US regulations by testing his pill in Puerto Rico, where his hundreds of research subjects received little information and were discounted as unreliable if they reported negative side effects. Rock's pill was unethically and imperfectly researched. It was marketed as "cycle control" to try to fool the Catholic Church—Rock basically presented it as a way to stabilize your period to improve your ability to avoid fertile days, even though it worked by preventing ovulation and implantation. It was difficult to get if you weren't married. It could cause fatal side effects. But by 1962, some 1.2 million American women were on it. Sex would never be the same.

To summarize what we've learned so far, people have pretty much always had a hankering for reliable birth control. But what about abortion?

SMASMORTION

Modern pharmaceutical birth control generally works by preventing ovulation, keeping sperm from successfully making it to the egg, or keeping a fertilized egg from implanting itself in the womb. But while many may act as if abortion is a modern invention, for much of human history it was an extremely effective method of fertility control. Anyone who's been paying attention will have noted the frequent overlap between contraceptives and abortifacients. Before the age of reliable pregnancy tests and a solid(ish) understanding of how fertility worked, there was a finer line between things you took just in case you might get pregnant and things you took just in case you *were* pregnant.

Aristotle and Plato both presented abortion as a good form of population control and generally preferable to killing or abandoning an infant at birth. We have reason to believe, by the way, that exposing an unwanted infant was quite widely accepted, given that Aristotle once commented on the noteworthy tendency of Egyptians to "rear all their children that are born."[32]

(It's right to be horrified that Greco-Roman culture so readily allowed families to discard living infants. Yet I think it's worth pointing out that in 2020—when a quarter of folks polled in the United States believed abortion should be illegal in most cases and politicians in several states were pushing to overturn *Roe v.*

Wade—6.1 million children lived in food-insecure households.[33] Some 4.4 million US children lived without health insurance that same year.[34] For a nation that many politicians would have us believe is "pro-life," we sure are leaving a lot of our kids in the cold.)

Old Testament Hebrews were likewise fairly pragmatic about pregnancy; the closest thing to a commentary on fetal rights appears in Exodus, which lays plain that if someone attacks a pregnant woman and she dies, they'll face the death penalty. If she only miscarries, they'll owe a fine to her husband—an indication that fetal death was seen more as a loss of potential than as a loss of life.[35] The Jewish Talmud further cements the idea that the rights of personhood are only granted upon birth; until a baby's head has emerged from the birth canal, Jewish law suggests that protecting the life and health of the mother should always come first.[36]

While Hindu and Buddhist laws both condemned abortion in the BCs, ancient medical texts from predominantly Hindu and Buddhist cultures also described how to perform one. In Myanmar, Thailand, Malaysia, Indonesia, and the Philippines, people wishing to terminate their pregnancies still sometimes turn to a painful and risky form of external womb massage that is depicted as either a punishment or a sin in Southeast Asian carvings dating back to the ninth century.[37]

Much of the modern debate around abortion is tied to Christianity. But even the church's relationship with the practice has always been complex. Vehement condemnations of this aspect of family planning arose as early as the first century, with some Christian leaders equating abortion with murder. But others seem to have disagreed, with allowances for pregnancy termination under certain circumstances. And historical texts tell us that even many of the folks who saw abortion as murder only felt as such after the "quickening"—when a person first felt the fetus inside them move—or otherwise after the fetus was "formed," meaning it took on a recognizable human shape. The abortion of an "unformed" fetus in the first trimester was more likely to be looked at as a minor transgression or even as an acceptable form of birth control than as the killing of an innocent soul.[38]

That was still the letter of the US law in the nineteenth century, which is why untrained "doctors" could peddle abortifacient pills and surgical abortions openly in newspapers in the early 1800s.[39] Using only slightly veiled euphemisms like "uterine tonics" to solve "female troubles," people like Ann Lohman (née Trow)— also known as either Madame Restell or the Wickedest Woman in New York, depending on whom you asked—got rich commercializing folk herbal remedies of varying efficacy. These services weren't solely utilized by ruined young women and ladies of the night; Restell notably had a sliding scale and charged wealthy clients up to a hundred dollars.

Now, I'm not going to stick my neck out and champion Restell as a feminist icon. We have no way of knowing how many of her remedies actually worked (lots of these mail-order docs sold toxic substances that put their patients in great peril and only sometimes caused abortion as a side effect) or whether she performed her surgical interventions responsibly. But while she and others may have been quacks, that's not why abortion was virtually illegal by 1900.

I am sorry to say that we must once again blame the patriarchy. In fact, we can pretty reasonably blame one guy in particular: Horatio Robinson Storer.[40]

It all started in 1847, when the American Medical Association (AMA) first formed. The concept of "medicine" as a uniform field with standards and shared best practices was pretty new. As such, nineteenth-century physicians were eager to distinguish themselves from—and overtake—the surgeons, midwives, and "doctors" who might just as soon sell you poison as give you reputable health counseling. Storer, a Boston-born and Harvard-educated obstetrician oft called the father of gynecology, was at the forefront of the effort to delegitimize the status quo for women's health care, which had always been sort of an inside job. But how to convince women to stop relying on female midwives for medical care? By demonizing some of their practices, for starters. Storer led the "physicians' crusade against abortion" starting in the 1850s, arguing that fetuses were people and that women were designed to carry them; any desire to do otherwise, he claimed, was a sign of insanity. He also thought that removing ovaries was a great way to cure mental illness, so, you know, there's that.

He seems to have thought abortion to be something of a trendy vice: "It has been said that misery loves companionship: this is nowhere more manifest than in the histories of criminal abortion. In very many instances, from our own experience, has a lady of acknowledged respectability, who had herself suffered abortion, induced it upon several of her friends: thus perhaps endeavoring to persuade an uneasy conscience, that, by making an act common, it becomes right."

Storer's personal beliefs on fetal personhood might have been genuine; he converted to Catholicism in want of a church that shared his convictions. But regardless of Storer's true motivations, the AMA's pivot to vehemently opposing abortion conveniently served to frame midwives—who had historically performed such procedures—as backward and evil. That couldn't have hurt business for fresh-faced new ob-gyns like Storer. Thanks to his efforts, the AMA stood against abortion until the 1960s.

That had a real impact. Between 1860 and 1880, at least forty antiabortion statutes were written into law, and many of them referenced the AMA's stance that life began at conception in their rationale. By 1900, virtually the whole country had outlawed the practice.

Even during the midst of this legal shift, abortion remained commonplace. Archaeologists recently found bones from a roughly thirty-week-old fetus in the nineteenth-century privy pit of an upstanding middle-class family that lived on Canal Street in New York City.[41] Records show the residents had several children, and the researchers also found an empty bottle of Clarke's Female Pills on the scene—a mail-order drug containing savin, a known abortifacient.

In the 1955 book quoted in the epigraph to this chapter, Hungarian French ethnologist and psychoanalyst George Devereux attempted to sort the attitudes around abortion in four hundred "primitive" societies into logical categories organized by motivation, method, and so forth.[42] (Weirdly enough, he claimed to do this to teach us about not abortion but about categories. I find this very baffling, but I will say he had cheekbones that could cut glass.) By virtue of the book's having been written by a white European in the 1950s, its anthropology citations create a sort of minefield of potential misunderstandings or blatant misrepresentations of various Indigenous peoples, which is why I'll refrain from listing specific cultures and their methods included therein.

But it's worth knowing that Devereux's academic colleagues had *dozens* of examples ready for him to analyze and sort. Abortion existed in all corners of the world for all manner of reasons. If reproduction is a natural act, then history tells us again and again that the drive to control reproduction is natural too. It's not just the results of sex we're desperate to control either—we're also obsessed with making a notoriously finicky act happen on demand.

Chapter Nine

WHY DON'T OUR BODIES ALWAYS COOPERATE WITH OUR HORNY HEARTS?

....

In which we do our darndest to make dicks get hard.

....

BY NOW, WE'VE FIRMLY—PERHAPS EVEN TURGIDLY—ESTABLISHED THAT sex is fundamentally essential to life, the universe, everything, and that its importance has colored our physical and psychological evolution and can be blamed for at least a solid third of our hang-ups and other nonsense. So it must be the simplest thing in the world, right? You just need two or more consenting parties to get the party started? Absolutely not. Sex is a physical pain in the butt for almost as many people as it is a mindless act of easy-breezy carnality.

When I was but a wee freshman in college (Go, Llamas!), I encountered the classic Mesopotamian text called the *Epic of*

Gilgamesh. It's largely considered to be the oldest surviving work of great literature, which makes its four-thousand-year-old, two-week-long sex scene some of the most ancient written smut in existence. Enkidu, an ill-fated wild man created for the sole purpose of interacting with the title character, spends a fortnight of his extremely short life getting his world rocked, continuously, by a priestess. (It was actually just a one-week-long sex scene, back in 2008, but someone unearthed a new page containing more sex at some point in the intervening years, which I didn't learn until fact-checking this manuscript.) In a truly classic sixteen-year-old-Rachel gag, I decided to write in to the school paper's advice column under the guise of being a young penis-haver intimidated by Enkidu's exploits. "I just feel so inadequate," I wrote. "What can I do to increase my stamina?"

The slightly older teens running said column had the frankly shocking foresight to answer the absurd question in all seriousness. At the time, I was annoyed that they didn't pick up the joke and run with it. But a decade and change later, I realize how important it was for them to take a teenage boy's anxiety about sexual performance at face value. Most people think of erectile dysfunction (ED) as an affliction affecting only the most aged gentlemen. But contrary to popular belief, even horny youngsters can have trouble enjoying sexual encounters. While research on teen sex is scant, one small study in France found that more than one in ten of its young male subjects had mild to moderate erectile dysfunction.[1] Nearly half of the female participants reported issues such as low libido, pain during sex, and difficulty orgasming.

Female sexual dysfunction is woefully understudied and underreported, with many women not even realizing their experiences with sex could be improved with medical or interpersonal intervention. Another nascent area of research is sexual dysfunction of transgender and nonbinary people; while an improved relationship with one's physical form and gender identity can *psychologically* pave the way for better sex, many typical hormone therapies and surgeries can create new physical issues, such as lowered libido or pain—or at least present a new learning curve that far too many medical professionals are unwilling or unable to help patients navigate.[2]

Point being: no matter what age you are or what's in your pants, having trouble getting it up (either literally or metaphorically) is perfectly normal. And folks have been trying to mitigate those troubles—and capitalize on our anxieties about them—since time immemorial.

Look, friends, not everything is pathological. Sometimes the answer to regaining a long-lost erection is simply to find some pretty people to kiss, try not to stress out too much, and give yourself time to figure out what floats your boat. Sometimes a lack of physical arousal means you just don't really want to have sex, which is always worth noticing and honoring. But that doesn't mean there's any shame in looking for shortcuts when your mind is determined to do the nasty and your body won't cooperate. So here are a few of the tricks—from goat testicles to penile injections—we've tried to turn each other on throughout history.

LIMPING ALONG

In his 1927 book *Impotence in the Male*, Austrian physician Wilhelm Stekel posited that inopportune flaccidity was a disorder "associated with modern civilization." As an avid follower of

Sigmund Freud and a man most preoccupied with the mental motivation behind various sexual hang-ups, Stekel unsurprisingly blamed limp members on the psychological shortcomings of the society he lived in. But Stekel couldn't have been more wrong. In fact, textual references to erectile dysfunction could date back as far as the eighth century BC.[3]

Misbehaving penises first reared their ugly heads (or failed to rear them, as the case may be) in the *Sushruta Samhita*, a Sanskrit text on medicine attributed to the great Indian surgeon Sushruta. The author of the text didn't land on one definitive cause for this unfortunate malady; instead, he suspected that various diseases of the genital organs or other systems might play a role but also blamed "the rising of bitter thoughts" and "forced intercourse with a disagreeable woman."

But that sketchy diagnosis didn't stop him from offering up possible remedies. One solution was to mix sesame paste and hog's lard with particular types of legumes and rice commonly used medicinally, season the mash with salt and sugarcane juice, and fry it up in clarified butter before consuming. This dish, Sushruta claimed, would allow any man to visit one hundred women before growing weary. Various testicles also came into play; one could poach goat testes in milk and consume them with sesame seeds and porpoise fat, or one could simply boil the sacks of alligators, mice, frogs, and sparrows in ghee before rubbing the oily infusion on the soles of one's feet.

Ancient Egyptians often assumed sexual dysfunction to be the result of a curse or hex.[4] A purely physical problem might be solved with complex cocktails of herbs and other natural substances: one ancient text recommends hyoscyamus, willow, juniper, acacia, ziziphus, myrrh, and yellow and red ocher, among other ingredients. We have no idea whether topical application of these poultices would have helped, but physically applying a tingly tincture to your genitals is bound to get *something* going. But if magical forces were suspected, these medicinal treatments went hand in hand with prayers and rituals. At least one remedy involved inscribing the name of one's worst enemy on a cake of meat and feeding it to a cat, according to historians. Metaphorically feeding your enemy to a sacred feline might not fix your underlying urological issues, but you can't blame folks for feeling a lot better afterward.

Any logic informing the sex-enhancing therapies prescribed by ancient physicians is long lost to us, leaving modern researchers to speculate about what various tinctures and potions were actually meant to do. But according to some research on the ancient Greeks, one possible mechanism for getting people hot and bothered was to get them warm and farty.

Second-century physician Galen, oft-cited as one of the most brilliant medical minds of his time, offered a multitude of suggestions for dudes hoping to stay hard. If a bull peed in the dirt and produced a small puddle of mud, rubbing it onto your penis as a plaster would imbue you with ferocious fecundity. Drinking a potion made from lizard kidneys could induce a fearsome erection. But in the works of Galen and his contemporaries, one commonality we see again and again is the inclusion of zesty ingredients like pepper and arugula. Because sexual arousal was (and still is) associated with feeling flushed and warm, it's easy to see why ancient physicians would consider spicy food a short-cut to bliss. But it wasn't just about spice: flatulence may also have been key.

Farting your way to a healthy erection makes sense if you think the way Galen did. See, he believed that, in addition to blood, human veins carried a "vital spirit" called pneuma: a kind of life-giving air to complement our more obviously present life-giving liquids. Galen believed that erections were the result of blood and pneuma rushing in to fill the arteries of the penis. He was at least half-right; a soft phallus is full of spongy tissue that becomes firm when filled with blood. But because Galen believed that air was also an important part of the equation, classicist Brent Arehart of the University of Cincinnati argues, he likely encouraged his male patients to create as much internal wind as possible to help power their sexual encounters.[5] With a belly full of gas and hot food, Galen figured, a man had everything he needed to go on a veritable spree of turgid thrusting. As anyone who's ever tried to turn Taco Tuesday into a romantic evening can attest, this advice, while no doubt well meaning, was misguided.

A LITTLE SOMETHING FOR THE LADIES

Sometime around the sixteenth century, in the bustling metropolis of Islamic medieval scholarship that was Timbuktu, someone stopped to think of the ladies. They penned a manuscript called *Mu'awana al-ikhwan fi mubshara al-niswan*, or *Advising Men on Sexual Engagement with Their Women*. This wasn't exactly a forward-thinking, feminist tome. The general idea was for male readers to use their newfound sexual prowess to gain back the favor of wayward wives. And plenty of its advice was geared toward the typical aphrodisiacal aim of increasing male vigor: suggestions for upping potency included drinking dried and pulverized bull testicles (a classic choice), fumigating oneself with the smoke of a burned nail from a rooster's right foot, and licking pulverized lizard penis mixed with honey. Prayer was also recommended for improving and prolonging sexual exploits.

But unlike most historical texts on improving sexual satisfaction, the Timbuktu manuscript also included methods to get female libidos going.[6] That doesn't mean the advice was *good*. One suggestion was that a man wipe "both his and his wife's eyebrows and hands with the gall bladder of a fox" (according to a paraphrased synopsis from researchers). Another tip: rub rooster blood on your penis before intercourse (don't do this). Looking to give your wife an orgasm "to the point of madness due to the intensity"? Just slaughter a black chicken, rub it all over your dick, and go to town (don't do this either).

THE BONER POLICE

Those embarrassed about occasional impotence would do well not to visit medieval Europe. (Actually, as I hope this book makes abundantly clear, no one should ever take *any* opportunity to visit medieval Europe.) A husband's inability to get it up was just about the only reason a woman could file for divorce or annulment at the time in England, since men incapable of fathering children weren't even supposed to get married.

Sex wasn't just forbidden outside marriage; even married couples were expected to limit their couplings to brief, straightforward, conception-centric affairs. Church documents from throughout the Middle Ages denounce the sinful lasciviousness of heterosexual married folks who

- had sex at a time of the month when conception was unlikely;
- did anything fancy and unnecessary to conception, like *insert any position but missionary*;
- had sex more than once every couple of weeks;
- used the pull-out method.

Given that strict sense of who and what and where sex was meant for, one can see the church's reasoning in condemning men incapable of penetration to a life without marital bliss. With marriage existing solely to create a set of circumstances under which two people could technically have a little tiny bit of a very specific kind of sex—a necessary evil in order to keep the religion from dying out—one couldn't simply have a wife and not impregnate her. Stop hogging all the fertile wombs, lads! You're not even using 'em!

From a modern standpoint, I'm tempted to applaud *any* opportunity a woman in the Middle Ages had for divorce; women in the Middle Ages generally had opportunities for nothing but dying in childbirth or being burned at the stake. But the circumstances were far from ideal for either the husband or the wife involved.

First, a woman had to wait at least three years to declare that she'd been hoodwinked into marriage to an impotent man. This window was presumably meant to allow for the possibility that a husband was simply shy, malnourished, ignorant as to the mechanics of marital sex, or overwhelmed by the long list of circumstances under which the church had assured him that he should *not* screw his wife. Second, the accusing spouse had to have character references from local townsfolk—yes, this was a completely public affair—because women are, as you know, inherently wicked and untrustworthy. Last, and most hilariously, the court had to receive proof that true impotence was afoot. In the twelfth century, a group of "wise matrons" would spend several nights hanging around the couple in question to accomplish this.[7] As a result, twelfth-century court documents are rife with testimonies from respectable married women on the complete inadequacies of various men's penises.[8]

But what could a man do to avoid losing his marriage and reputation? If he had access to an actual scholar, he might receive advice from the tenth-century tomes of Tunisian physician Ibn al-Jazzar, widely considered about as good as it got for the whole span of the Middle Ages. His main shtick was to insist that testicles be kept warm and moist, which he said could be accomplished by eating foods such as chickpeas, turnips, ginger, long peppers, and beans (note: hot and farty).

For the majority of these spurned husbands, however, the bulk of the blame would be put on witchcraft, with suggestions running the gamut from asking nicely for the witch to stop hexing your genitals to catching her in a choke hold.

There were also a few slightly less murder-y solutions available. In the thirteenth century, Friar Albertus Magnus noted that the frenzied copulation seen in sparrows made their meat a perfect cure for frigidity. A roasted wolf's penis might be even more potent, but the most salacious meal of all was a starfish, which he warned might work *so* well that a patient would ejaculate blood.

Never fear: these violent emissions could be cured with a "cooling" meal such as *checks notes* a plate of lettuce.[9]

THE G.O.A.T.

It says quite a bit about our preoccupation with sexual prowess that one of the most infamous medical hucksters in US history peddled goat testicles.

Born in North Carolina in 1885, John Romulus Brinkley wasn't *really* a doctor: a series of unfortunate events kept him from ever managing to finish (or pay for) medical school, so he bought a degree from a diploma mill instead.[10] Still, Brinkley is said to have made millions of dollars curing, in his own words, "sexual weakness, insanity," and a host of other conditions. His method of choice was, to say the least, quite novel: He claimed to graft the virile testes of healthy male goats into the scrotums of tens of thousands of patients, thereby reinvigorating them sexually and allowing tired, sterile men to become energetic fathers. Later, after years of tussling with the newborn American Medical Association and even butting heads with the Federal Radio Commission for being a sort of turn-of-the-century Dr. Ruth and running one of the most powerful transmission towers of the time, he claimed that his famous goat-grafting surgery had inspired a simple injectable formula that could cure literally all ailments with the power of . . . goat balls.

If you're not yet grasping that Brinkley was a quack to end all quacks, consider that he also sold anticancer toothpaste.[11]

When he finally faced justice for his flimflammery (which, by the by, only happened because *Brinkley* sued one of his naysayers in the medical world for libel, which failed spectacularly and opened a floodgate of aggrieved patients and bereaved relatives ready to demand reimbursement), Brinkley was forced to admit that his "testicle transplants" weren't transplants at all. While many of his happiest patients assumed they'd been given powerful new nuts, the surgery actually only involved slipping a sliver of goat tissue *between* the scrotum and the skin. Without attachment to a blood supply, one expert witness in the libel suit testified, the grafted tissue was merely a foreign object akin to a splinter—and would either dissolve or form a scar. Far from infusing his patients with superpowered ungulate sexuality, Brinkley was merely exposing them to a risk of infection or rejection in the hope of inducing a dramatic placebo effect. And the formula he replaced those sham surgeries with? An independent chemical analysis revealed it to be one thousand parts distilled water mixed with one part blue dye. He sold the injections at a hundred bucks a pop, and each patient received an average of five.

Brinkley may be the most infamous testicle salesman in history, but he wasn't alone. Serge Abrahamovitch Voronoff spent the roaring twenties and dirty thirties experimenting with xenotransplantation in France and landed on a specialty of stuffing monkey testicles into human scrotums.[12] It's worth noting that Voronoff was likely merely misguided where Brinkley was unscrupulous; by all accounts he did at least insert his grafted tissues into his patients' actual scrotums. Based on his animal experiments, Voronoff was convinced that grafting a younger animal's testes onto patients could reinvigorate them (in fact, he'd wanted to use human death row criminals as his donors but wound up starting a monkey farm to source from instead). His procedures fell out of favor—probably no thanks to the likes of Brinkley—but similar investigations into the power of testes persisted.

When Dutch scientists finally isolated the hormone testosterone in 1935, it briefly seemed as if the work of Voronoff and others would be vindicated. Here was a substance testicles emitted, providing a possible mechanism by which grafting on a new one might put some pep in your step. Alas, while a boost of testosterone does increase libido in many male patients, research has failed to show it imbues the general youthfulness and vitality promised by gland-grafting physicians. We still have a lot to learn about this hormone, which is an essential part of various bodily functions in all sexes, but it's certainly not a panacea for anyone with a penis.

A HEALTHY GLOW

Feeling less than frisky? In the early 1900s, you might have tried popping some radium pills. Discovered by Marie Skłodowska-Curie and her boo Pierre in 1898, the highly radioactive element radium seemed to many to have endlessly promising commercial applications. You could use it to make slippers that glowed in the dark! And clock hands that glowed in the dark! And glass that glowed in the dark! Wow.

Whether due to the sheer newness of the stuff, the fact that it gave off a glow, or both, folks got it in their heads that radium was the key to perking up tired bodies and curing all manner of ailments. You might drink slightly radioactive water to treat your gout or take radium salts to improve your arthritis. All of these products were advertised in newspapers and available for purchase by mail. And like any commercialized cure-all, radium was also marketed as a solution for low libido.

Many supplements slyly promised "vigor" or "vitality" (you can't tell me that's not supposed to be about screwing, y'all). But others got straight to the point: Nu-Man and Vagatone gland tablets promised to restore "sexual power" and "remedy sexual apathy," respectively.[13] You might also try a Radioendocrinator or an Adrenoray: these radium-packed products were placed directly on your endocrine or adrenal glands to juice them up with sexual strength. Got a vaginal infection? Maybe some hemorrhoids? Vita Radium Suppositories could fix whatever was wrong with wherever you shoved 'em and make you horny to boot. The Testone Radium Appliance and Suspensory was particularly special; the user would tuck their testicles into a little silk pouch full of radium salts, then tie bands of fabric around their waist and legs to keep it all in place. Always a good idea to make sure you keep your balls radioactive while you're on the go. And if you wanted to make your sex organs feel warm, plump, and tingly instead of cold, clammy, and lifeless—any sex organs, all types, wow—you could caress those frigid bits with Magik Radium Massage balm.

Please don't put anything radioactive on your junk.

THE LOGICAL BASSOON

Giles Brindley is undeniably a man of many and varied talents. In the 1960s, the UK native developed a neuroprosthesis capable of restoring some sight to the blind and casually invented an instrument he dubbed the "logical bassoon." According to a 2014 profile published in the *British Journal of Neurosurgery*, he spent his sixties taking up marathons and relay racing; as this book went to print, he was in his nineties and studying the origins of falsetto.[14]

Brindley is a polymath if ever there was one. But if he wanted to be most remembered for his life-altering work in prosthetics, his sexagenarian sportsmanship, or his endeavor to create a more perfect bassoon—well, he shouldn't have flashed a room full of people in Vegas.[15]

Picture it: Nevada, 1983. Glitz. Glamour. The annual meeting of the American Urodynamics Society. The world's top urethral whizzes are gathered to discuss the latest in bits and bobs, testes and taints, and urination and ejaculation. Some of those in attendance claim that when Brindley took the stage to present his work, he faced a particularly sophisticated audience. His was the final talk scheduled for the day, so many of his colleagues were already dressed for a swanky evening reception. A few of the bigwigs had reportedly even brought their dates along, dressed in party dresses and gowns.

It's important that you picture a room full of folks dressed to the nines and desperate to hit the conference reception's open bar. Can you see them? Great. Okay. Because now you need to picture Brindley shuffling onstage in a pair of track pants and proceeding to show that crowd a slideshow of his penis in various stages of tumescence.

Brindley was not having some kind of hypersexual breakdown. In fact, the unassuming Brit had spent some years previous injecting himself with substances he thought might induce an erection: a concept that was pretty much unheard of in the mainstream medical community at the time. Erectile dysfunction was generally considered kind of a *you* problem and perhaps something to discuss with a psychoanalyst. The promise of anything more physically invasive had been tainted by the past failures of goat-grafting hucksters and monkey-farming dreamers or else required *serious* surgical intervention with a marginal payoff. In the 1970s and 1980s, for example, hundreds of thousands of men had inflatable silicon rods inserted into their genitals, creating more reliable erections at the cost of hosting fallible equipment and frequent infections in their nether regions.

But Brindley's work on neuroprosthetics had brought him a lot closer than most to solving the problem of flaccidity. In developing a sacral nerve stimulator to help improve bladder control and mobility in people with spinal cord injuries, Brindley also came to study—and implement—techniques for making paralyzed men ejaculate.[16] Sexual performance problems among the able-bodied may have been considered a mostly mental issue. But for paralyzed men unable to have children due to their inability to get erections, the problem was clearly clinical—and clearly worth solving, at least in Brindley's mind. In one 1984 paper, he confirmed nine healthy children as the result of his latest round of trials with vibrators and electrostimulation.

Shocking a penis into spilling its guts is one thing, of course, and inducing a serviceable erection for the sake of pleasurable intercourse is another. But Brindley saw the possibilities.

By injecting himself with a relaxant called phentolamine—just one of many substances he tried to varying degrees of success (read: rigidity)—Brindley managed to loosen the walls of involuntary smooth muscle that make up human blood vessels. In a penis, relaxed arterial walls mean more room for fluid. That makes blood rush right in, which springs the penis into action.

Brindley's unexpected slideshow demonstrated these effects. But given the long-standing belief that erections were mostly a mental game, he saw fit to prove he hadn't secretly fluffed himself up for those glamour shots off camera. Who cared what you'd injected if you might have been sexually aroused at the time anyhow? Brindley assured the crowd that *no one* could be turned on by the prospect of giving a urology lecture (which definitely isn't true, because everything under the sun is arousing to *someone*, but we'll let that slide for another chapter or two) and informed them that he'd injected himself with the good stuff just a few minutes prior. That's what the track pants were for: Brindley stepped away from the podium to let the thin fabric show off the fruit of his labor in his Fruit of the Looms.

What happened next is shrouded in sensationalism and hearsay. Some would have you believe that Brindley boldly dropped trou. Others say he sheepishly dropped trou. There are numerous stories of coiffed ladies shrieking in horror as Brindley stepped down from the stage, approaching his audience erection first, while other accounts of the evening suggest that colleagues calmly requested a closer look (or even feel) out of professional curiosity. Whether the result was general pandemonium, a decorous scientific discussion, or some combination of the two, the facts we know are these: Brindley pulled his pants down onstage to display a magnificently tumescent penis and, in so doing, changed the field of urology forever.

Self-injection swept the world as a groundbreaking solution to erectile dysfunction and paved the way for the now ubiquitous blue pills known as Viagra. Brindley is responsible for getting countless people hard over the decades, and his work will no doubt continue inspiring erections long after he's gone.

⊼

Viagra was not supposed to give you a boner. It was supposed to treat angina, a cardiac condition otherwise known as ischemic chest pain.[17] (Fun fact: poppers, which provide such an intense sexual experience by dilating blood vessels and relaxing involuntary muscle that they've become a ubiquitous part of gay culture, were originally used to treat angina in the Victorian era.[18]) Viagra, which in its generic form is called sildenafil, blocks an enzyme called PDE5 in a bid to expand blood vessels and increase blood flow.

But when Pfizer conducted clinical trials for sildenafil, they found that the compound left the body so quickly that patients would have to take it multiple times a day to see a real cardiovascular improvement. Taken with such frequency, it had the downside of causing muscle aches.

Fortunately for Pfizer, it also caused boners.

In time—with more disappointing results on the angina front and more data on side-effect erections—Pfizer decided to pivot to testing the drug specifically as a treatment for ED. For reasons unknown (cough, cough, sexism, cough), they didn't choose to pursue its effects as a treatment for uterine complaints—we now know it can help treat menstrual cramps.[19] Pfizer went full steam ahead to bonerville.

The rest, as they say, was history. Viagra hit the market in 1998, and it's now one of the most popular drugs in the United

States. A 2015 study found that the US Department of Defense spent $84 million on ED aids for members of the military in 2014 and more than $41 million on Viagra alone. (Not-so-fun fact: Health care for gender-affirming medications and surgical procedures would only cost the military an estimated $3 million to $8 million a year, but such expenses are far more controversial.[20]) Now people with penises can seek erectile dysfunction drugs online, without the potential embarrassment of a doctor's office or pharmacy visit.

But what if you don't have a penis? The hunt for so-called female Viagra continues apace. Addyi (flibanserin), which was initially intended as an antidepressant but fizzled in clinical trials, came out as a treatment for low sexual desire in women in 2015.[21] But while Viagra works in a pretty straightforward way (blood vessels dilate, more blood flows through them, more blood flows to penis, penis gets hard), Addyi works on the brain. It's also taken daily, which stands in sharp contrast to popping a Viagra about an hour before you want to go to pound town. Taking a drug that can cause low blood pressure, nausea, and fainting every day on the off chance that it might tweak your brain chemistry—a notoriously tricky thing—in such a way as to make you enjoy sex more is, frankly, kind of cuckoo bananas. After pretty much bombing for all the reasons I just listed, Addyi had a flashy relaunch in 2018 and became easily and cheaply accessible online.[22]

In 2019, the Food and Drug Administration approved Vyleesi (bremelanotide) as another "pink pill" option.[23] On the plus side, it's taken on a similar schedule to Viagra; you just use it right before you want things to get spicy. It also has fewer side effects than Addyi. But it still works on brain chemistry, so its efficacy is bound to vary widely—plus it's not a "pink pill" at all but an autoinjector syringe.

So, for now, there really aren't awesome pharmaceutical options out there for people with vaginas who have trouble feeling as physically frisky as they'd like. You might do better to talk to a therapist about how you and your partner(s) might change up what you're doing to better flip your "on" switch or to reconsider if and when you actually *want* to be having sex at all.

It's okay not to want to have sex *ever* or for it to be more of an interesting or funny or ambivalent physical act that you enjoy for more cerebral or emotional reasons. It's also normal and, while frustrating, *totally okay* not to physically be able to have sex whenever you wish you could.

Whatever genitals you have, don't assume there's something wrong with you when they don't work in lockstep with your brain. While you can and should seek solutions if a lack of libido is negatively affecting your life, try to seek them holistically. Find partners who are patient with you and want to communicate with you about your needs. Try different ways of having sex (news flash: no one has to penetrate anyone to have a good time), and try to separate your desire for intimacy and pleasure from what you think sex "should" look like.

Some people rarely, if ever, have orgasms but have plenty of fun building toward them. Some people rarely, or never, experience sexual desire but still enjoy having physical intimacy like kissing and cuddling with a partner. Some people are up for having sex five times a day and fall in love with people who quite literally aren't built like that. Some people don't want physical intimacy at all and have to learn for themselves what emotional partnership, should they want it, looks like for them.

All of this is okay. If the only thing standing between you and a magical evening is a penis that will get hard on command, by all means try Viagra or something similar (with a doctor's consult, of course). But don't be surprised if things are a little more complex than that—and don't be afraid to explore what it is you might need. Especially if all you need is a little porn.

Chapter Ten

WHAT IS PORN, EXACTLY?

· ·

In which we fail to define porn and accept that it is everywhere.

· ·

LET'S REVIEW WHAT WE'VE LEARNED SO FAR. HUMANS HAVE ALWAYS had lots of sex. What's more, they've frequently obsessed over how to do it differently, how to do it better, and whether they should be doing it at all. Being preoccupied with the sex you're having (and the sex you're not having) is one of the most human pastimes in the world.

And for nearly as long as humans have had sex, they've found ways to depict it.

In 2008, archaeologists digging up a cave near Stuttgart, Germany, unearthed a sculpture estimated to be about thirty-five thousand years old; it's one of the oldest confirmed sculptures in the world (as opposed to some even-older "figurines" that might just be, uh, slightly bootylicious rocks). And what did this historic piece of artwork depict? A lady whose ass was bodacious (good gracious).[1] Several such shapely idols have been discovered at different points and places in prehistory, hinting that some of our species' first artisans did a thrifty business in carving boobs and butts.

But does the presence of a naked, fleshy lady mean the creator had pornographic intentions? My Instagram feed would argue otherwise (though Instagram's obscenity policies would say yes, absolutely, block that tasteful nude posthaste, 'tis surely a porn). Indeed, without the ability to ask our Paleolithic ancestors to weigh in on how their culture viewed these objects, we'd be revealing much more about our own hang-ups than theirs by assuming enlarged breasts and exaggerated sexual organs were meant to titillate.

It's just as possible that these works and others—carvings and cave paintings depicting giant phalluses and vulvas, which also date back to when we made our earliest scribbles—were more about encouraging fertility, or honoring the human form, or worshipping some extremely stacked goddesses in a totally platonic way.[2] We simply cannot know the intended purposes of these boobies.

In a 1964 Supreme Court opinion on obscenity, Justice Potter Stewart famously defined the bounds of hard-core pornography with the comment "I know it when I see it." Before we dive headfirst into the history of porn, we must pause to accept that Stewart's definition, while obviously easy to mock, is still about as good as it gets. Porn is not rigidly demarcated by certain types of images or sounds; porn is a category we created to separate stuff that feels like it's too much about sex for general consumption from all the other art and entertainment in the world. It's less a type of thing than a vibe—and a recently invented vibe at that.

See, as those ancient German tits so elegantly highlighted for us, we can't pinpoint the earliest evidence of "porn" without agreeing on what counts as porn. I'm partial to *Merriam-Webster*'s definition, which classifies pornography as any material—written, drawn, filmed, whatever—that depicts erotic behavior with the intention of triggering arousal. A picture of an erect penis is not inherently pornographic; it might exist in the context of a medical textbook or, like, a really sad-sack self-portrait. "I know it when I see it" resonates so well with our concept of pornography because it is not the substance of the stuff that makes it porn; it's the way it's presented to us and the way we consume it. Until the moment you see porn and know it as porn, it exists in a liminal state somewhere between an anatomical study, a potentially erotic artistic depiction, and a deposit for the spank bank.

Not to put too fine a point on it, but the word "pornography" as we know it only even dates to the nineteenth century. Karl Otfried Müller, a German classicist who focused on Greece, coined the word *pornographie*, inspired by the Greek *pornographos*, to refer to "obscene" relics he found in the course of his study of the ancient world. But the original Greek referred quite specifically to someone who wrote about *pornai*, or prostitutes. Müller coopted the term to mean "yuck, ew, get this stuff out of here" and applied it liberally to ancient pieces of art that offended the sensibilities of his day. The *Oxford English Dictionary* didn't include an anglicized version until 1909.

I'm really, really sorry to get semantic about porn here, but the point I'm spiraling down to is this: unless we have historical context that assures us people were making and viewing a particular piece of art with arousal in mind, we can never really be sure it was pornography as we now know it.

Why does this matter? Because if we instead defined pornography as anything that *could possibly* be used to inspire arousal in the viewer, we could say, with certainty, that humans have been making porn for as long as they've been making *anything*. There are very few times and places in human history where depictions of some sort of sexual organs or sexual acts weren't commonplace. What's really interesting is that sometimes—and only sometimes—these materials have been deemed obscene. And maybe—just maybe—that means we should be a little less uptight about the way we consume media that depicts sex today.

As is generally the case when we talk about prehistory versus history, things got at least a little bit less ambiguous when the Mesopotamians hit the scene. While their erotic clay plaques dating to around four thousand years ago may not have been made exclusively for people to wank over—they've been found in the ruins and remains of temples, graves, and private homes—they are indisputably racy. In contrast to the curvy statuettes of our earliest ancestors, these popular tablets portrayed explicit acts of intercourse in all manner of heterosexual positions. And they had visual gags too: one particularly famous plaque shows a man and woman stopping to hydrate while in the throes of passion. Some scholars speculate that the image, which shows a bent-over woman sipping from a straw while the man penetrating her lifts a cup of wine to his lips, is meant to make the viewer think of oral sex.

Not every ancient culture churned out nude figures and lewd scenes. Ancient Egyptian temples and tombs are notably prim and prudish in their depiction of couples.[3] But searching for erotic art on Egyptian walls may be the equivalent of looking for a copy of *Girls Gone Wild* at a public library: just because the smut wasn't put proudly on display doesn't mean people weren't making it.

Despite the general gentility of the vast majority of preserved art from Egyptian antiquity, researchers know of at least one approximately three-thousand-year-old papyrus so dirty in nature as to be dubbed "the world's first men's mag."[4] It features a series of vignettes of attractive young women engaged in explicit acts with enormously well-endowed men—blokes who, according to a 1993 analysis by art historian Gay Robins, didn't fit the mainstream aesthetic of the upper class at all. Given that, and the other material in what's now known as the Turin Erotic Papyrus (anthropomorphized animals doing mundane tasks), this old-timey nudie mag seems less like a serious piece of hard-core porn and more like the sort of material middle school boys might giggle over. Still, Robins and others note, the quality of the illustrations suggests that whatever the intended *purpose* of the publication, its intended *audience* was classy. So, while we might not know much about how ancient Egyptians did or didn't use racy images to get their rocks off, we do suspect that at least some members of polite society got a hoot out of raunchy literature.

Since we can't, as a society, seem to agree on what porn is and we can't know what ancient folks actually used as masturbatory fodder, I can't trace you a straight line from the lewd frescoes of the ancients to the explicit Tumblr GIFs of my own youth. Humans have simply drawn boobs in too many ways throughout history for me to weave a unified theory of smut. Although we can point to examples of potentially erotic objects across a huge span of history, we have much less sense of how they figured into society.

But there just might be a lesson in that. Perhaps that ambiguity is a reminder that sex—wanting it, not wanting it, having it, and not having it—is so tied to who we are and how we live that treating depictions of it as belonging to some magical, weird category that's separate from all other art and media is . . . unhelpful. No, I'm not saying that small children should be allowed to watch hard-core porn on their iPads. But maybe the reason why they *shouldn't* isn't all that different from why it's not great for them to see airbrushed models in magazines: it's not realistic and might confuse them while their brains are still all nice and mushy. And we do a real disservice, I think, in lumping together all images of nude bodies as *erotic*. Our modern concept of pornography too often casts sex as inherently dirty, inherently obscene—something not discussed in polite company—and the human body itself as potentially pornographic. The archaeological record hints at a different way of thinking in our species' past.

A TOTAL SEX POT

Most archaeologists and anthropologists now agree that to call the fabled sex pots of Moche "pornographic" is reductive, if not completely incorrect. But they are, indeed, sex pots, and like many items discussed in this chapter, they shocked the colonizers who first spotted them out of context.

(Just pause briefly to imagine famed Egyptologist Howard Carter holding a candle up to a tiny opening in King Tut's elusive tomb, with

his landed gentry patrons behind him in their Victorian best. They ask, "Can you see anything?" and he answers, "Yes, wonderful smut.")

The Moche or Mochica were a people so named for what is thought to have been their bustling capital city on the northern coast of modern-day Peru. They lived from around the second through the eighth century AD and saw their peak from around 300 to 600.

We don't know a ton about Moche civilization, because its people didn't leave written records and their particular society was long gone by the time Spanish colonizers arrived in Peru around 1526. In fact, the Moche seem to have been subsumed by the Chimú, a culture conquered in turn by the Inca in the mid-fifteenth century. But in addition to several temples, an irrigation canal system, and signs of human sacrifice, the Moche people left behind several hundred explicit sex pots. Yes, sex pots.[5]

I don't want to suggest that Moche sex pots are the only reason we should care about this culture's ceramics. The thousands of pitchers and cups and urns found since the Spanish arrived show evidence of the use of molds to mass-produce vessels *and* a tendency to decorate them with realistic imagery. If you wanted to drink out of a handmade cup with your face on it in pre-Columbian South America, it seems, Moche was the place to go.

But Moche is now most infamous for the spouted vessels that Catholic colonizers reportedly smashed in horror upon their discovery (and of which we've luckily uncovered hundreds more since). For reasons still unknown, the Moche decorated some of their ceramics with graphic sex scenes—most of which featured anal or oral in favor of vaginal intercourse and many of which

included nursing infants on the breast of the female participant. The religious colonizers of the sixteenth century very much
considered all of this no good, and the relics continue to push
the boundaries of modern sensibilities, though the Spaniards
wouldn't likely have been much more amenable to photorealistic
depictions of penis-in-vagina marital sex anyhow.

Suffice it to say that historians and anthropologists have put
forth dozens of potential explanations for the creation of these
vessels, which were once thought to be strictly funereal but have
since been found in residential areas as well as tombs. Perhaps
the Moche understood that breastfeeding can lower fertility and
included it in images of nonprocreative sex to encourage family
planning. Or maybe, some have suggested, the Moche believed
that the afterlife was a world of opposites: perhaps burying
loved ones with pots that depicted nonprocreative sex would
allow them to have children in death. It's also possible that the
pictured scenes reveal some part of a belief system around fertility that we're not familiar with; the Moche would not be the
first culture to believe that something other than vaginal intercourse was required to trigger a healthy pregnancy. The fact that
some vessels include a living participant dallying with a skeleton
could support either of the latter two theories, perhaps signaling
that, at least in some cases, these pots were meant to affect a dead
person's sex life or indicating that connecting with one's ancestors
was more crucial to continuing your family line than the physical
act of sex with a living partner could be.[6]

It's also possible, of course, that the Moche just liked creating and viewing tableaus of sex and that the tendency toward
anal and oral simply reflected tastes of the day (or aspirations, at

least). The frequent inclusion of infants may have been meant to broadcast the general fecundity of the people pictured or to signal that this wasn't the female partner's first rodeo. Maybe the idea was that anal sex improved the quality of a mother's breast milk. Weirder things have been believed about sex and reproduction much more recently in history.

All of the scholarly discourse on the Moche sex pots stands in sharp contrast to how Spanish colonizers reacted to them in the mid-1500s, which was to do a big smashy-smashy and to use the artifacts as evidence of the depravity and baseness of the people whose land they were stealing. Even setting aside the obvious travesty of colonization in general, this framing seems especially unfair given that the Moche pots were around one thousand years old and Europeans had gotten up to all sorts of weird shit in their own recent history.

But porn is in the eye of the beholder, and the Spanish swore they'd known it the moment they saw it.

MISGUIDED CENSORSHIP

Japan is not without its own taboos around sex. Homosexuality remains a matter of social and religious controversy, and Japanese obscenity laws are infamously intense: in 2014, an artist was arrested for sending code that could be translated into a 3D model of her vagina for at-home printing.[7] But it's also a place where you can buy used underwear in vending machines.[8]

This could be what happens when you try to hastily tidy up for Western visitors and their delicate sensibilities in the face of thousands of years of widespread acceptance of erotica. Men in power tried to do just that during the Meiji period at the turn of the twentieth century. Historians argue that occasional attempts to squash explicit artwork before then had more to do with politics than a genuine cultural distaste for smut; a brief period of censorship in the 1700s is thought to have been a reaction to some particularly pointed political comics that happened to feature sex.

Because Shinto and Buddhism, the prevalent religions in Japan, both present sex as natural, the Japanese have long viewed it through a fairly neutral lens.[9] Shinto in particular features stories of gods engaging in sex and uses phallic symbols to promote health and fertility.

Our incapacity to define porn is never clearer than when we discuss *shunga*, a style of erotic art that saw its heyday in Japan between the seventeenth and nineteenth centuries. When the British Museum in London hosted an exhibit of the genre, most famous for Katsushika Hokusai's *The Dream of the Fisherman's Wife*, curator Timothy Clark told the BBC, "People who haven't seen shunga before will be surprised by how explicit it can be. But this is sexually explicit art, not pornography, produced to exactly the same technical perfection as art in other formats by the same people."[10] In the very same article, he told the reporter that, well, actually, maybe they were sexually explicit art *and* porn: "The division between art and obscene pornography is a Western conception," said Clark. "There was no sense in Japan that sex or sexual pleasure was sinful." If you have to think of something as sinful for it to be porn, I haven't seen porn since I was fourteen years old. Point being: if your culture enjoys erotica as mainstream art, you'll be beset with outsiders who either think everything you make is porn or that nothing you make is porn.

The weirdness with which Japanese erotica has traversed the ages—being mainstream art, then being hurriedly swept under the rug, then scrabbling its way back to being a major cultural phenomenon—has definitely had its downsides. For starters, Japan only outright outlawed child pornography in 2014.[11] In *2014*. From the 1980s up until that point, it was an appallingly popular way to get around censorship laws, which prevented showing pubic hair but had no explicit rules against displaying unclothed children. Cartoons, animations, and games are still allowed to portray child porn.[12]

I don't know if there's a lesson in that. But if there is one, I think it's that people are going to make smut no matter what you do, so you really shouldn't make the biggest loophole in your obscenity laws include child abuse.

CAN YOU KEEP A SECRET CABINET?

You'll be forgiven for assuming that hard-core porn is a recent invention—or at least something that only recently became popular. After all, most of us didn't grow up hearing tell of erotica in our school history books or spotting smut on the walls of natural history or classical art museums.

That's no accident. Much of the potentially pornographic material produced by our forebears has been more than forgotten; it's been deliberately hidden away to protect the modern public. I'm not going to try to convince you of some grand historical conspiracy to make us believe humans haven't always loved porn. Let's just put it this way: if I had a nickel for every "secret museum" that existed solely to keep ancient porn away from prying eyes, I'd have several nickels. Like, not a *ton* of nickels. But it's weird that I'd have more than one, right?

One of the most telling examples of what happens when historical erotic art meets the immovable object that is a prudish society took place in the early nineteenth century, when King Francis I of Naples decided to take in some culture with his family at the national archaeological museum. By the time he visited in 1819, the facility was home to a range of artifacts from Pompeii and Herculaneum. These Roman cities had been destroyed by Mount Vesuvius more than seventeen hundred years before and only excavated in earnest in the few preceding decades.

We don't know exactly how Francis and his family fared at the museum that day. But he was supposedly so shocked that he ordered several pieces to be secreted away behind locked doors to protect his subjects' delicate sensibilities. Whether or not this origin story is totally accurate, we know that the most explicit pieces of art and decor from these lost cities did indeed get removed from the main floor. The so-called Gabinetto Segreto, or secret cabinet—a reference to "curio cabinets," not a comment on the size of the space, which featured far too many phalluses to fit in a single pantry—spent the better part of a century and a half away from public view, though it was briefly opened to the masses during periods of peak liberalism such as the 1960s.[13]

On the one hand, it's not surprising that nineteenth-century viewers were a little taken aback by the art of ancient Rome. Frescoes and sculptures depicted explicit scenes of intercourse, showcased massive sets of genitalia, and depicted gods, goddesses, and other mythological figures as powerfully sexual entities.[14] But the sheer number of penises and breasts on display in the ruins of Pompeii and Herculaneum suggests more than just a love of porn. Erotic art was simply a fact of everyday life in a way difficult for us—let alone our nineteenth-century counterparts—to imagine. (In other words, yes, they knew it when they saw it.)

For example, the ruins of Pompeii held a wind chime in the shape of a giant, flying bronze phallus. Giant members, with or without men attached to them, were apparently such common decorative features on household objects that it's highly unlikely people saw them as particularly titillating. If one person in a small town made a wind chime out of their favorite silicone dildos and butt plugs, their neighbors might think displaying a clearly sexual object to the public was a bit kinky. The owner wouldn't necessarily be *getting off* on their unique decor, but the act would reasonably be seen as subversive—a declaration of an unabashed enjoyment of masturbation—and the artiste surely would expect it to appall at least *one* neighbor. If dildo wind chimes took off on Etsy, on the other hand, and suddenly became a feature on most porch fronts, we'd recognize that the decor held some meaning beyond sexual titillation. Such was the way in Pompeii, where historians suspect that comically large (and sometimes winged) members were talismans to promote fertility, agricultural bounty, and wealth.

That's not to say all the art King Francis took issue with was purely symbolic; the Gabinetto Segreto also featured lewd frescoes from all over the doomed cities—public areas, bathhouses, and brothels—that showed various configurations of couples in joyful congresses of all sorts.

One piece of Pompeii and Herculaneum's history is particularly infamous. The sculpture known as "Pan copulating with a goat" is exactly what it says on the tin, but the goat seems, like, really into it. Which is admittedly a little uncomfy.

Speaking of goats: guess where else you *couldn't* find a statue of a satyr screwing an ungulate in the mid-1800s. That would be the British Museum, *unless* you were an individual of "mature years and sound morals," in which case you could *absolutely* gaze upon a satyr copulating with a goat in statue form.

The London institution's Secretum opened in 1865 in response to the Obscene Publications Act of 1857. Fun fact: That piece of legislation came about when the chief justice, Lord Campbell, oversaw a trial related to the sale of pornography at the same time as he participated in a House of Lords debate on the sale of poison. The sale of those two items—which I really hope we can agree are, like, not the same at all—really got Campbell's brain juices flowing, and he famously declared that smut peddlers were engaged in "a sale of poison more deadly than prussic acid, strichnine or arsenic." His solution was to introduce a bill that outlawed the sale of porn outright, though he very reasonably targeted "obscenity" without defining what counted as obscene. (Why does this keep happening?)

Selling porn was actually *already* a misdemeanor at this time—hence the trial that gave Campbell the bright idea to further criminalize erotic materials—but the Obscene Publications Act made it a statutory offense, giving courts the power to seize and destroy any obscene materials they found for sale. Again, we must pause to reflect on the fact that *no one defined what counted as obscene.*

Given the ambiguity of this antierotic legislation, it's completely understandable that scholarly institutions took pains to make clear that any nudes in their possession were of the tasteful sort—and that anything on the racy side was kept out of view. Unless rich dudes wanted to see it. So, in some ways, the definition of porn in Victorian England had a lot to do with the social status of the person looking at it.

According to the British Museum, upon its creation the Secretum contained around two hundred objects labeled as "abominable monuments to human licentiousness."[15] A mature and moral gentleman could enter this smutty sanctum to view, along with the aforementioned goat sculpture, examples of Pompeii's phallus fixation, old-timey condoms, and sexual scenes plucked from the walls of an Indian temple, among other things that probably weren't actually *porn*, so to speak.

As of the printing of a 2011 guidebook, the Gabinetto Segreto was just another art gallery—albeit one that kids under the age of eleven were ostensibly forbidden from entering. The British Secretum had an even less climactic decline, with curators gradually spreading its objects out across the museum starting around the 1960s. The goat, however, remains conspicuously absent from modern displays.

But while Victorian sensibilities may have balked at the idea of free and easy access to erotica, these folks were nowhere *near* as prudish in their private lives.

FROM DAGUERREOTYPES TO *DEEP THROAT*

My first encounter with porn came at the age of nine, and the porn was exceedingly bad. I stumbled across a user-generated choose-your-own-adventure game online. It started with basic scenarios—you walk out your front door and decide to (a) take a left, (b) take a right, (c) go back inside, and so on—but because anyone could add their own selections for actions and got to write the stories for what those actions led to, the whole thing quickly devolved into raucous (and typo-ridden) sexual escapades.

I was fascinated. I was titillated. Spending hours clicking through the erotic tales became an obsession. When I finally got the guts to disclose my new fixation to a slightly older friend, she was like, "Uh, have you heard of fan fiction?" Let me tell you, it was all over for me after that.

I was not alone in my descent into internet erotica, and the options extend far beyond user-generated text. After all, my discovery of lascivious content came just a couple of years before the first reference to "Rule 34": if something exists, there's porn about it somewhere online. Just as bumbling discussions of sex were common enough for me to accidentally find them as a child, drawings, photos, and videos of sexual acts were (and are) common enough that those in the know could track down whatever might tickle their fancy at any time of day. From blurry amateur endeavors to high-production-value celebrity streams, the age of the internet means that porn of any sort is always just a click away.

That level of accessibility has, according to many advocacy groups, turned us all into pornography addicts. The National Center on Sexual Exploitation (founded by clergymen in 1962 and previously known as Morality in Media), for example, has lambasted everyone from HBO to *Cosmopolitan* for glamorizing immoral and dangerous behavior; it sees the situation as so dire that it wants the American Library Association to censor public internet use in libraries. And, well, consumption has indeed gone up: a 2015 study in the *Journal of Sex Research* found a 17 percent increase between the 1970s and the 2000s in young men who reported having viewed an X-rated film in the previous year.[16] Young women went up by about eight percentage points.

Whether that uptick is good, bad, or neutral is a complicated question. But while our digital lifestyles have made porn more *accessible* (and, by extension, a bit more *acceptable*), our appreciation for erotic words and images is far from new. Humans have been harnessing technology to make and distribute erotica for the entire history of technology.

Consider, if you will, the Victorian era. Picture it. Are you picturing it? Decorum. Ladies in high-necked dresses and stays. Men with well-groomed mustaches. Propriety. Toodle pip!

Now consider, if you will, Victorian erotica. Orgies. Gay sex. Cunnilingus. Dildos, like, everywhere. Lots of pee.

The oft-shared impression of Victorian-era society as being incredibly buttoned-up (both literally and figuratively) is not false: wealthy members of society adhered to strict standards of politeness and propriety when in public, and women were expected to stay virtuous (and often ignorant) until their wedding nights. But many Victorian men and women were *obsessed* with sex, and their desire for titillating content knew no bounds.

Just as the internet has allowed twenty-first-century humans to expand their appetites for pornography, Victorian-era technology created a booming new industry for smut. Louis Daguerre invented the daguerreotype, a precursor to photographs as we know them, in 1839; someone had started circulating a daguerreotype depicting penetrative sex by the 1840s. Videos of sex became de rigueur by the turn of the twentieth century, though they were made clandestinely for the first few decades and were pretty much exclusively passed around in brothels and other male spaces. These so-called stag films were short and to the point; they generally featured ten minutes or so of silent, vigorous sex and nothing else.

Things took a turn in the 1960s and 1970s, when artists like Andy Warhol tried their hand at erotic filmmaking and amateurish films like *Deep Throat* managed to pull off wide theatrical releases. *Deep Throat* is far from high art (and according to its star, Linda Boreman, it contains footage of her being assaulted and acting under duress, so you can just take my word for it—it's not worth a watch). But it stood in sharp contrast to the kinds of porn that had been circulating up until that point, because it at least *attempted* comedy and sleek production. It wasn't made to be viewed in dark corners by men with their hats pulled low; it was made to be seen in plush theater seats with friends.

Something about the culture of the moment made people excited to see and talk about this and other newly minted pornographic films in a way that hadn't previously been deemed socially acceptable. In a 1973 *New York Times* article, Ralph Blumenthal noted that the film had "become a premier topic of cocktail-party and dinner-table conversation in Manhattan drawing rooms, Long Island beach cottages and ski country A-frames" and had drawn couples on dates and even single women to theaters across the nation. "It has, in short, engendered a kind of porno chic," Blumenthal concluded.[17]

Sure, the Supreme Court acted that same year to button up obscenity laws enough to keep explicit films down to limited theatrical releases. But for a brief moment, going to see sex for sex's sake on the big screen was a perfectly acceptable thing to do with your buddies (at least in some circles).

Once VHS tapes proliferated in the 1980s, porn became easy to watch in the privacy of your own home—which meant it was way less awkward to masturbate to. Now we've circled back to those soundless stag films in the form of explicit porn GIFs, which can deliver you your thrust or money shot of choice on a loop for infinity with no convoluted plots to waste time. The more things change, amiright?

Chapter Eleven

WHY DO SO MANY OF US LIKE SEX THAT ISN'T "NORMAL"?

....

In which people get freaky.

....

READER, A CONFESSION: I JUST SPENT ROUGHLY TEN MINUTES STARING out my window trying to remember when I first became aware of the concept of kinkiness. At what point in my impressionable youth did someone let slip that sex could get weird? I truly cannot recall. Unlike so many topics in this scribbled tome, I have no formative childhood incident around which I can build a unifying theory of what-we-talk-about-when-we-talk-about-fetish-gear.

However.

I *do* remember discussions of perverts. Deviants. Dirty sex fiends who were twisted in the noggin. My first encounter with the concept of a perv is crystal clear to me; I'm not sure how young I was, but it was young enough that my Bible school teacher found six ways to Sunday to avoid telling us what anal sex was when she taught us about Sodom and Gomorrah.

For those unfamiliar with ye olde S&G, these twin cities of sin show up in the Old Testament Book of Genesis, when the Lord tells Father Abraham that their residents are very wicked indeed and will be punished with hellfire. Abe, the mensch that he is, asks if God might reconsider if the mortal can prove there are fifty righteous people in Sodom (this number drops as time goes on to demonstrate just how depraved Abraham finds the Sodomites to be and to show how generous and merciful God is, which is all very relative in the context of the Old Testament). Also important to note: there were two angels present for this misadventure, for reasons I cannot recall and which are not particularly relevant.

Then comes the really juicy part: While lodging with Abraham's nephew Lot, the holy travelers are accosted by a growing group of men from the city. Translations vary here, but it's generally accepted that the good townsfolk of Sodom wanted free rein to bang the angels. Lot, being a good host, is like "No, you cannot rape my angelic guests," but, being an Old Testament guy, he also is like "Would you consider raping my two virgin daughters instead?" The crowd says no (which makes one wonder if God would have preferred they say . . . yes?), the heavens rain brimstone on both Sodom and Gomorrah (which seems unfair, given that this neighboring urban area seems to have only been guilty by association), and the righteous Abrahamic characters flee (except for Lot's wife, who gets a little sad about her town being on fire and is, naturally, turned into a pillar of salt as a result).

Somehow, the generally accepted Christian takeaway from this story, at least in the circles I grew up in, is that God did not like that the men of Sodom wanted to have sex with boy angels. Many modern scholars of these texts have pointed out that asking permission to rape someone's houseguest is bad form regardless of the sexes or genders involved in the equation, not to mention the fact that the parable seems to dwell just as much on the urban area's lack of support for the poor and its residents' rampant consumerism as it does on whatever sex the angels did or didn't have. So maybe that wasn't supposed to be the part people zeroed in on as evil. Those angels probably weren't "boys" anyway, as such celestial beings are canonically genderless, though that nuance always got glossed over in my Sunday school. Also, it's really worth pausing to ask what the holy heck was up with Lot offering up his daughters.

But despite recent evidence to the contrary, I am not here to give you a slapdash lesson in theology. I'll get to my point.

When I, as a wide-eyed Sunday school attendee, was introduced to the concept of deviance—of sex perverts so nasty they got themselves fully smote—my teachers didn't *have* to go into detail about what kind of lewd nonsense would constitute such heights of perversion. They didn't even really focus on the threat of assault. Sodom and Gomorrah were simply presented to me as places where men liked sex. Like, a lot. The implication being, of course, that they liked it more than a "normal" amount. Also, sometimes with other boys. The ladies of S&G never really came up, so I don't know if we were supposed to imagine that female horniness and acts of random lesbianism were also to blame for God's wrath.

Before the notion that people might enjoy whips and chains or feet or smooshing cake on one another in the bedroom ever crossed my pretty little head, sexual deviance had a much simpler definition: people who were so into screwing that it made God uncomfy.

If you're reading this book, I assume you don't think that "enjoying sex" or "being attracted to androgynous beings from literal heaven" counts as "weird." But while the standard Christian pervification of Sodom and Gomorrah is just one extreme example, the inclusion of same-sex attraction and general horniness on lists of pathological perversions was widely accepted until quite recently. Only in the last few decades have the scientific and medical views on deviance shifted *slightly* away from "anything but a cis-man and cis-woman doing missionary in a monogamous relationship a number of times per month that doesn't strike us as odd" to focus more on practices that are, at least arguably, truly fringe.

But now there's more and more data available on what gets people around the world going. And there are more and more opportunities for folks to find community around and give voice to the desires they might once have thought themselves alone in having. As such, it's becoming apparent that more of us deviate from the mean of sexual normalcy than our general discourse around sex would imply.

Hard-core studies on being hard-core in the sack are hard to come by. But in one survey of some one thousand folks from a swath of demographics in Quebec, researchers tallied that around a third of respondents had experience with voyeurism or exhibitionism; more than a quarter recounted indulging in some kind of fetishism.[1] According to the researchers who conducted the study, the threshold below which a behavior can be considered "statistically unusual" is around 16 percent. Having 20 percent of respondents claim experience with masochism might not *sound* like a lot, but that means if you get together ten random folks for a party, there's a good chance that two of them will be able to argue about who needs to service top.

All of this raises a question: What's actually normal, anyway?

A KINK BY ANY OTHER NAME

Let's agree on some terminology before we go any further. As I pondered on in the introduction to this chapter, there's a wide gulf between what we folks today typically think of as "deviant" or, in more intimate lingo, "kinky" and what some dominant school of thought might deem "perverse" at some time or place in history or culture. This semantic disconnect is the kind of problem we're bound to run into when we try to create a universal language by which to codify and describe all the ranges of human behavior.

But there's also the simpler issue of most modern people just throwing around words, without (1) really knowing or caring or (2) stopping to agree upon what they mean. I'm going to be fumbling with the first problem throughout this whole chapter, and you may have noticed me fumbling with it for several chapters already. But we can settle the second problem right here and now. Here are some of the words we'll need to share an understanding of if we want to get through this subject. You also might find them useful the next time you get into an intellectual debate at a sex dungeon.

PERVERSION

1. A perverted form; especially an aberrant sexual practice or interest, especially when habitual.

Example sentence: Depending on whom you ask, wanting to have sex for reasons other than conceiving a child can be considered a sexual perversion.

PERVERT

1. One who has been perverted; (specifically) one given to some form of sexual perversion.

Example sentence: Depending on whom you ask, wanting to have sex with your spouse can make you a pervert!

KINK

1. A short, tight twist or curl caused by a doubling or winding of something upon itself.

Example sentence: Peter's kink is getting slapped with the pointy end of a garden hose kink.

2. A mental or physical peculiarity; eccentricity, quirk.

Example sentence: Judy's got a thing for that guy's weird kinks.

3. A clever, unusual way of doing something.

Example sentence: "What a great kink," she thought as she stared at the new knot she'd invented.

4. A cramp in some part of the body.

Example sentence: He loved sitting in stress positions until the kink in his neck became unbearable.

5. An imperfection likely to cause difficulties in the operation of something.

Example sentence: Your unwillingness to have freaky sex with me is sure putting a kink in this relationship!

6. Unconventional sexual taste or behavior.

Example sentence: Peter's kink is getting slapped with the pointy end of a garden hose kink.

FETISH

1. An object or bodily part whose real or fantasied presence is psychologically necessary for sexual gratification and that is an object of fixation to the extent that it may interfere with complete sexual expression.

Example sentence: Barry's boot fetish turned out not to be very compatible with Joe's thing for getting his toes nibbled on.

2. Fixation.

Example sentence: Am I developing a fetish for holding your hand, or is this what love is like?

PARAPHILIA

1. A pattern of recurring sexually arousing mental imagery or behavior that involves unusual and especially socially unacceptable sexual practices (such as sadism or pedophilia).

Example sentence: Mark wanted to talk to his therapist about his paraphilia but got reasonably freaked out by the fact that Merriam-Webster *lumped* sadism *and* pedophilia *together so casually.*

PARAPHILIC DISORDERS

1. Paraphilias that cause distress or cause problems functioning in the person with the paraphilia or that harm or may harm another person.

Example sentence: Once he found a kink-aware therapist, Mark realized his fantasies, while unusual, weren't paraphilic disorders because they didn't bother him or motivate him to cause harm to other people.

WHENCE CAME KINK?

What do we now consider kinky? And where did said kinkiness originate?

First off: here, "we" means, I guess, most people in America. But, full disclosure, my brain may be skewed a little left of center on this one because I'm not *boring*.

When we look to the origins of kink, we run into a lot of the same problems we muddled through in our previous chapter, when searching anthropological archives for pornography. We can't assume a fresco of, say, a man being led around on a leash had the same kinky edge at the time it was painted as it does today. Maybe putting your man on a leash was simply the thing to do at the time, and calling such behavior kinky would be akin to suggesting that consenting young adults who sext each other today are engaging in something truly taboo. Or perhaps the leash served a practical purpose or was part of some kind of ceremony that had nothing to do with sex: one that honored familial bonds or harkened back to some canine creation myth. (This is, to be clear, all entirely made up. Please don't cite this in an online roundup article about freaky stuff people did in the past and start a rumor about a Mesopotamian sex cult that revolved around dogs.)

What's "normal" comes and goes in waves and varies from place to place. And so too must what is, by contrast, perverse.

Don't worry, we can still have fun trying to define this mess. But we'll be answering two separate questions here: What have different cultures in time and space seemingly accepted as main-stream behavior that would be considered freaky-deaky sex stuff by most modern folks? And what practices did they *actually* deem weird?

SEEING SEX WHERE IT ISN'T

In thinking about the broad history of sex, it's important to re-alize just how much our sense of what sex should or shouldn't be skews our perception of the world. And I don't just mean how it skews our perception of *sex*. Our ongoing need to draw straight and thick lines around what sex should and shouldn't look like can also lead us to see sex where it isn't.

Consider, for instance, the genital preparations traditionally undertaken on the infants of Hawaiian islanders. According to notes from the early colonial impositions on the islands—and from twentieth-century anthropologists—it was once commonplace for female relatives to blow air into the foreskins of infant males. This was done to loosen the foreskin over time, which would ease the subincision procedure most boys endured around age six or seven. While both the ethics and safety of partially splitting a young boy's penis are dubious at best, there is no reason to believe there was any sexual intent behind the air-blowing process. Indeed, this and other genital-preparation techniques— baby girls reportedly had their vulvas massaged with breast milk and oil, and infants of both sexes might have their buttocks massaged to make them rounder—were treated as rote chores and part of basic infant care. The aim was to help the child grow into an adult worthy of sexual desire and capable of as much sexual pleasure as possible, but the practices themselves were no more sexual than a diaper change. Unsurprisingly, though, it's said that Christian foreigners were completely horrified by what they interpreted as pedophilia.

Similar squick arises from the mostly out-of-fashion tradition of a rabbi using his mouth to remove the blood from a newly circumcised baby, with news reports circulating every few years when the practice results in a deadly infection. Again, this practice is often not *good* for a baby: too many old men are running around Brooklyn with cold sores. But it's not *sexual*, and we should probably question why the thought of a mouth going near a baby's urethra is often treated as such. There's nothing sexual about an infant's genitals unless we impose adult sexuality upon them.

Then there are the aspects of historical life that were *decidedly* sexual, but still not as lewd as they sound outside their proper context. Until the Victorian era, for instance, in many European cultures people commonly shared beds. Like, any and all people. If someone stopped by your place on their way through town, they were probably going to spend the night in bed with you. You may recall a version of this practice (and the means people devised to avoid getting too handsy when under the covers together) from our chapter on courtship and dating.

But consider the implications this had for when people *did* have sex. Children shared beds with their parents. Servants might all sleep in one room. Until the Victorian era split houses up into many discrete chambers, privacy just wasn't really a thing if you weren't rich.[2] If one wanted to have sex with their partner at night and in a bed, they probably did so with someone else sleeping (or at least putting on a good show of ignoring them) nearby. Did exhibitionism even exist before we all got our own bedrooms? Debatable. (But it probably did, because there's always *somewhere* you're not supposed to have sex, and someone will always want to do it there.) Relatedly, having a quick tumble in an alley or a barn—something that now carries an air of deviance to it, because *sex is for doing in a bed or at least in a private space*—might once have been the best choice for the truly prudish, as doing the deed in your bedroom at night meant a far greater likelihood of being observed.

PERVERSIONS AND PARAPHILIAS

The ever-shifting definition of what it takes to be a certified freak is by no means limited to the annals of our species' distant history. As is the case with so many things relating to sex, our current idea of what is "normal" in terms of sexual desire largely dates back to the turn of the twentieth century. Things have changed radically since then in some ways. But we've still been taking baby steps toward nonconformity within the same basic framework that psychiatrists set in the early 1900s.

The *Psychopathia Sexualis*, first published in 1886 but popularized and republished into the next century, did not paint a kind picture of the "perversions" it detailed. The author, Dr. Richard von Krafft-Ebing, took the typical stance of denouncing any sexuality he saw as abnormal, including homosexuality. The fact that the case studies collected in the book (42 at first but 238 by the time he published his twelfth and final edition) came from his time working at a mental asylum didn't help soften this framing. And as a forensic psychologist, Krafft-Ebing was particularly interested in the sexual proclivities of people who had committed crimes related to their desires.[3]

What's rather hilarious about *Psychopathia Sexualis* is how the general public took it. How did they take it, you ask? They took it and ran with it. The book was the first deep dive on sexual pathology designed to help forensic scientists, doctors, and eugenicists understand whence perversion came and how they might treat it. But it was also the first deep dive on pervs most people had seen published, full stop. Krafft-Ebing reportedly added more technical jargon into his later editions and even translated more salacious passages into Latin, in an attempt to keep the public from devouring his academic treatise like porn. Some scholars have suggested that *Psychopathia Sexualis*, far from further stigmatizing the already stigmatized, actually created a sense of community by showing queer and otherwise nonconforming people that they weren't alone. Whoops!

When sexologist John Money popularized the term "paraphilia" starting around the 1970s—defining it as "a sexuoerotic embellishment of, or alternative to the official, ideological norm"—he was trying to create a framework for discussing deviance in a neutral fashion. Money was not an altogether positive figure in the world of sex research; he was the driving influence behind an infamous case of forced sex reassignment in an infant with damaged genitalia (discussed in Chapter 3) and allegedly imposed sexually abusive "treatments" on the subject and their sibling, likely contributing to both of their deaths by suicide as adults. His methods were unforgivable.

But Money keeps cropping up in medical texts, however uncomfortable his presence may be, because he *did* have some ideas about how sex, gender, and desire might work that helped us shake off a few of our Puritan cobwebs. He's largely credited, for instance, with introducing "sexual orientation" as an alternative to "sexual preference." In so doing, Money helped to solidify the concept that being some flavor of not-heterosexual was more than an avoidable phase or, worse, a condition that might be treated and cured.

Money and his contemporaries helped get the *Diagnostic and Statistical Manual of Mental Disorders* (*DSM*) a little bit closer to recognizing "weird" sex as normal (though he did argue that paraphilias should be categorized as mental illnesses, a judgment that has persisted since his time). Before 1980's *DSM-III*, this medical tome had referred to "sexual deviations," a term that included such disparate inclinations as homosexuality and pedophilia and lumped general sadism in with rape and mutilation. It's hard to believe that being a man who wanted to have vanilla sex with another man was once considered comparable to being a man who wanted to, like, eat women's faces, or whatever. But such is the baffling history of psychiatry.

Since then, we've seen a series of small tweaks that have moved away from pathologizing sex for its own sake. The 1987 update to the *DSM*, for instance, replaced "transvestism" with "transvestic fetishism." So, instead of implying that anyone who cross-dressed or wanted their partners to do so had a diagnosable disorder, the implication became that only people with an unhealthy fixation on such things was inherently unwell. Of course, what constituted an acceptable level of fetish remained pretty ambiguous, and such phrasing likely still marginalized some otherwise happy and healthy consenting adults. A 2000 update to the *DSM-IV* took some steps to address this, adding a second criterion for diagnosis: the paraphilia had to "cause clinically significant distress or impairment in social, occupational, or other important areas of functioning." An additional note for exhibitionism, frotteurism, and pedophilia stated that acting on the urges at all was inherently unacceptable; acting on sadistic urges with anyone but a consenting partner counted as pathological as well. The message was finally at least sort of reasonable: there's nothing wrong with getting off on slapping people across the face as long as you don't just go around slapping people all willy-nilly. Stick to mutually beneficial face slapping, and you're absolutely fine.

The *DSM-5*, which came out in 2013, further solidified the distinction between an offbeat sexual desire and a psychiatric condition. There is now a diagnostic distinction between a paraphilia and a paraphilic disorder; the paraphilia is the thing most of us would call a kink or fetish, and the disorder arises if liking that thing causes you distress, impairs your ability to function, or is inherently tied to causing harm to yourself or others. By "harm," here, we of course mean *actual* harm—abuse and mistreatment of another being—and not utterly destroying a partner in a safe, sane, and consensual manner.

The *DSM* now also readily acknowledges that there are countless paraphilias not listed in its pages. Some experts argue that the inclusion of some specific kinks and not others could imply a value judgment against certain proclivities. If that's true, then it's time for psychiatric texts to simply state that any sexual urge that distresses the patient or causes harm to others is diagnosable and leave it at that. The powers that be behind the *DSM* argue that the eight specific paraphilic disorders they chose to define—voyeuristic disorder, exhibitionistic disorder, frotteuristic disorder, sexual masochism disorder, sexual sadism disorder, pedophilic disorder, fetishistic disorder, and transvestic disorder—are relatively common and can have forensic importance in criminal cases and so are worth highlighting to ease diagnosis.

I'll pause here to say that no sexual impulse you might have is inherently evil. *However*, certain kinks, fetishes, fixations, and other deviations from "normality" obviously can't be put into practice *without harming someone*. When, say, a famous comic gets caught whipping his dick out in front of unsuspecting colleagues (totally hypothetical example here), he might go on the record as bemoaning his unfortunate and difficult-to-control exhibitionist kink. But what your lizard brain tells your junk it should enjoy is never an excuse to harm other people. And using other people to get your rocks off without their enthusiastic consent *is* harmful.

That doesn't mean there aren't sexual inclinations that cause genuine anguish for the people who have them. For example, researchers have only just begun to study the population of people with pedophilic urges who do not wish to act on them, but being in such a position clearly causes intense inner turmoil.[4] Therapy can help, but people with these inclinations may rightfully worry that a mental health professional will treat them poorly or contact legal authorities, even in the absence of an intention to act.[5]

If you feel sexual urges that tempt you to do harm, I can't tell you exactly what you should do or why you have these urges in the first place. But I can promise you that all human beings sometimes feel impulses to do things that they know are morally wrong, and having these urges—even for things that are extremely stigmatized for good reason—does not make you evil or irredeemable. I urge you to seek out professional support in managing your impulses, which I sincerely hope will relieve you of what I can only imagine to be a massive emotional burden.

In the meantime, consider setting rules for yourself to keep you from situations where you might feel an impulse to hurt someone. Maybe you can't be alone with teenagers. Maybe you can't drink with fans in hotel rooms anymore. There's nothing wrong with making choices that minimize the risk of harm to you and to people around you. If the rules themselves cause harm to other people—let's say, totally hypothetically, that you're a politician who refuses to take meetings with female colleagues lest they drive you to temptation—then you really need to get yourself help ASAP. But there's no shame in that, so long as you seek out the support you need.

For now, we know very little about where our unique sexual proclivities come from. Unfortunately, that means there's no clear-cut way to manage one that causes pain. But hiding from the reality of your desires isn't the answer.

WHAT'S UP WITH THE WHOLE FOOT THING?

While we're on the subject of whence came the concept of perversion, you might also be wondering where more pedestrian kinks come from. Whether you find yourself in a tizzy every time you see Ben Solo put someone in a Force choke hold or you have a desperate need to lick armpits, it's natural to wonder why certain things that turn you on are benign—or even repugnant—to others.

Unfortunately, we don't yet have enough data for someone to create a unified theory of yucks and yums or to pen the treatise *On the Origin of Kink*. But science has a few intriguing thoughts about why some humans wind up attracted to things that seem unusual to others.

One fetish with a particularly neat-and-tidy (though far from proven) scientific explanation is having a fixation on feet. Also known as podophilia, a foot fetish is just one example of something called partialism: where you tend to fixate on a particular body part above all others. Some forms of partialism are accepted as fairly normal (a heterosexual man being nuts over boobs or butts on ladies, for instance), while others (feet) are considered weird but are still quite common.

I seem to be a magnet for men who are really into noses (?!), so my perspective on this may be slightly off. But my anecdotal take is that being into a particular body part is downright pedestrian, when all possible body parts are taken as a collective. But it makes sense that fixation on body parts that are more closely associated with traditional sex and physical sex characteristics is considered more normal than a focus on parts that are generally not part of the sex act, and feet—low and figuratively or literally dirty as they are—are particularly polarizing.

According to one 2006 study by researchers at the University of Bologna, University of L'Aquila, and Stockholm University, which analyzed several hundred internet chats about kink involving thousands of participants, partialism of some kind is by far the most common fetish. Feet and toes stand out as the most commonly fetishized body parts, when more traditional areas like bums are excluded. Another study found that the desire to sniff smelly feet or socks ranked as the most common form of fetish for objects associated with the body. Even Idris Elba, indisputably one of the sexiest actors of our era, has openly confessed to having a thing for women's feet.

Anyway, people love those little piggies. Like, an extremely not-insignificant number of people. Perhaps you are in their number. What's with all the foot fetishes?

As promised, foot fetishism has a uniquely tidy potential explanation. Neuroscientist V. S. Ramachandran studies the way crossed wires in the brain can cause amputees to experience phantom limb syndrome. Ramachandran offers one reason that feet might be a common target of ardor: the part of the brain that receives signals of podiatric touch is right next to the part that receives genital stimulation.[6] "Maybe even many of us so-called normal people have a bit of cross-wiring," he wrote in his 1999 book *Phantoms in the Brain: Probing the Mysteries of the Human Mind*, "which would explain why we like to have our toes sucked."

That link, while satisfyingly simple, is far from achieving scientific consensus. And because many kinks and fetishes don't deal with body parts at all, we certainly can't attribute *all* of our weirder inclinations to unusual neurological wiring.

Sigmund Freud viewed perversion—like he did just about everything—as likely stemming from periods of disorder in childhood development. He also thought that foot fetishists probably grew up equating feet with penises, which is just too much of a stretch for this gal to buy into.

Most of Freud's theories are now considered by many to be overly simplistic at best and sexist hogwash most of the rest of the time. Still, there is some truth to the notion that our most fundamental sexual urges may be formed in childhood. Mostly based on case studies and anecdotal wisdom, the idea is quite widely accepted that a feeling of proto-sexual excitement or attraction felt as a child can lead folks to sort of imprint on some arbitrary aspect of the stimulating scenario. I have heard psychiatrists suppose that as children our sheer proximity to feet—the easiest body part to see when we are low to the ground—may leave us prone to fixating on them as a visual when we register some new and intriguing feeling.

In truth there are likely as many roads to developing a fetish as there are things to fetishize, if not more. There may be some evolutionary component, perhaps because a diversity of sexual preferences and desires made people more likely to go forth and multiply than if we all wanted and competed for the exact same thing.[7] If so, it would be reasonable to assume that some of us are simply more likely, on a genetic level, to be open to freaky stuff.

Some kinks may be the result of social pressure. On the one hand, we might see TV and film suggest a particular thing is desirable, which influences our development; on the other hand, we might even be pushing back against what we're told is normal or civilized simply because taboo things are thrilling. It's also possible that sensations or objects that are easily swapped in the brain, as may be the case with genitals and feet, are a common source of seemingly bizarre sexual proclivities.

Many people with kinks likely develop them, at least to some extent, long before they consciously think about sexual activity, let alone engage in it. And this leaves plenty of opportunity for random experiences and thoughts to cement aspects of our sexuality that we find unusual—or even alarming.

Sometimes the fantasies we want to see play out on a page or a screen are a shockingly far cry from things we'd like to experience in reality. One obvious and unfortunately stigmatized example is the "rape fantasy," which more than 60 percent of female participants in one 2009 study reported having at some time or another.[8]

The term itself is reductive, even based on that one survey of some three hundred or so women. Nearly half of them reported that the fantasies created a mix of erotic and aversive feelings, and the study didn't further distinguish between nuances. There's a big difference, for instance, between imagining someone you find attractive and *wish* you could have sex with just absolutely losing control and ravishing you (quite a common daydream, if bodice rippers are any indication) and thinking about a stranger or distasteful acquaintance violently assaulting you. Both of them, to be clear, are okay fantasies to have, but there's quite a spectrum of potential scenarios in there. That's why a growing number of kinksters and kink-aware researchers use the term "consensual nonconsent" (CNC) as a better umbrella for all the ways you might very reasonably want someone to throw you onto a bed and have their wicked way with you.[9]

It's also totally normal for people who don't identify as female to have fantasies like this, by the way. One survey of four thousand Americans found that CNC musings were only slightly less common in self-identified men than in self-identified women (54 versus 61 percent, though men also reported having them less often). And they were even higher in people who identified as nonbinary (68 percent).

I obviously can't speak for every soul in existence, but I'm going to go out on a limb and say, statistically speaking, pretty much no one actually wants to endure sexual assault. So why do so many of us like fantasizing about it, or at least scenarios adjacent to it?

No one has a definite answer. As with other kinks, perhaps the sheer wrongness of it excites us, and that excitement sometimes translates into actual arousal. For some people, this desire may reflect an impulse to give up control, especially if they're anxious about initiating sex themselves due to trauma, social conditioning, or low self-esteem. It might sometimes be a way of reprocessing and taking control of an episode of actual violence in one's past. In any case, being into CNC isn't anything to be afraid or ashamed of, though the *consensual* in there is, as always, extremely crucial.

It's important that you ask yourself a lot of questions about what you like about these fantasies and why before you try to play around with them in real life. If your desires and boundaries aren't clear, you run the risk of an unscrupulous dickwad pushing your limits and hurting you. This can happen even despite your best efforts, and it's not your *fault* if it does. But you can save yourself a lot of potential pain by thoroughly exploring your desires *solo* before ever bringing in another player. If you like being on the aggressive side of CNC scenes and you don't spend pretty much every second thinking about making sure your partner is 100 percent comfortable and consenting, kindly go back to the kink shallow end until you're ready for adult swim.

Don't fret too much if you feel confused by the sudden appearance of a seemingly off-putting fantasy. There have actually been times when huge swaths of communities have collectively gotten titillated by scenarios they almost certainly didn't actually want to experience. One example: Jewish kids and Nazi porn.

In the early 1960s, Israel was packed to the gills with smut about Nazis. Just *chock darn full* of the stuff. Pulp novels called stalags that featured torture and rape in German prison camps were the top reading materials of eighteen-year-old Israeli boys in 1963, according to a Hebrew University survey.[10] The books were presented as translations of English works, though we now know they were produced by and for Israelis. They generally featured gentiles from the American or British forces being assaulted by buxom female SS agents until they broke free and exacted revenge in the form of . . . more rape.[11]

Israel was notoriously hard on pornography, but this pulp fiction managed to spread like wildfire. The final straw seems to have been when a book hit newsstands that featured not just a female protagonist but a Jewish one. That brought the taboo too close to home, and the genre was banned; examples of it were largely destroyed, and a small selection was secreted away to a hidden collection in the national library for the sake of historical record keeping.[12] The 2008 documentary *Stalags* argues that the genre flourished in a culture of silence around the Holocaust, which left the children of many survivors grappling with complex trauma when high-ranking Nazi Adolf Eichmann's trial was televised in 1961 and laid many of the horrors of the war bare for the first time.

We like this story not because of the scantily clad Nazis (ew). Instead, this story is important because it reminds us that getting turned on by things that would horrify you in other contexts is normal. And sometimes, when all the freaky stars align, the young adult male population of an entire country can decide to normalize something so surprising that we still find it confusing and uncomfortable more than half a century later.

WHO'S YOUR LEATHER DADDY

Over the last decade or so, a debate has surged every spring about whether LGBTQIA+ Pride Month—which takes place in June in the United States—should center on, or even allow the presence of, people who outwardly display kinky tendencies. A person might obviously engage in kink in loads of ways, queerly or otherwise, but the quintessential out-and-proud display at Pride is that of the leather community. I'm not going to take a stand on whether assless chaps have a place in the parade; as a bisexual cis-woman in a relationship with a cis-man, I'll show up at any queer gathering that welcomes me with a smile on my face and a plate of cookies and mind my own damn business. (More power to you if you're in my demographic and feel more at home taking up space in queer places. As long as you recognize and work to address your straight-passing privilege, I sincerely hope you feel welcome and validated.)

Anyway, we're here not to argue about kink at Pride but to talk about how it got there in the first place.

The first explicitly gay leather bar didn't open in the United States until 1958, but leather culture had been simmering ever since the end of World War II. Motorcycle clubs popped up as a place for men to bask in military-adjacent hypermasculinity while eschewing mainstream culture. These subversive spaces attracted people who didn't feel they had a place elsewhere. Soon there were clubs—and then entire biker bars—composed of gay men. Being a biker was a great way to hang out with fellow queers: the subculture's signature activity normalized the idea of traveling in packs, while its style allowed for dressing in an offbeat way that showed off your body. In the 1970s, lesbians began to form kink-centric leather communities of their own.[13]

Look at it this way: People who identify as straight and cis live in a society where they are the default. That doesn't mean each one of them perfectly fits the mold; if you're a straight cis-woman who doesn't want or can't have kids, for example, you have to deal with a lot of nonsense due to the fact that most culture today revolves around the formation of a nuclear family. It's kind of like being white and Christian in America. You might not think you have a cultural identity around being white and having grown up going to church. But, actually, you also have constant access to Christmas and Easter paraphernalia and the ability to order the kind of food your mom made in even the most Podunk little town. White, nominally Christian culture (as opposed to, say, Ukrainian culture or Scottish culture or any of the other things white Americans often give up to fully embrace whiteness and its privilege) is boring as hell, but it's always been there for those Americans to bask in. They don't notice it because it's treated as the default setting.

Heterosexual culture exists too. It's what most people call "normal." It's dating and getting married to someone of the opposite sex and gender, and both of you having chromosomes and genitals that don't make anyone uncomfortable or confused, and combining your finances, and having two to three children. The further you get from being a straight cis-person, the less accessible that particular culture becomes, especially when you look back further than a decade or two. There's pain in that, but there's also freedom. And the men and women and nonbinary people who created the first leather communities were harnessing that freedom to create new spaces for themselves. Being gay wasn't acceptable in America for most of the twentieth century, full stop. Trans people were arrested for "cross-dressing" across the United States into (and no doubt, in some places, long after) the Stonewall Riots in 1969. So why would out-and-proud queer people have bothered to try to emulate straight culture, when they were among friends?

However you identify now or in the future, I hope this book has made clear that our concept of what is "normal" when it comes to sex and love and romance and family making and pegging your boyfriend is *really* arbitrary. You can take it or leave it. Normality just wasn't built to serve most of us. If there's something you want to do or think about or ask of your future partners, and it doesn't inflict harm or undue emotional stress on another person, but you *still* feel like it might just be too . . . *much*, ask yourself who and what made you feel that way. Ask yourself what they stood to gain by making you feel that way. Then go be gay and do crime.

AM I WEIRD? AM I NORMAL? AM I *TOO* NORMAL? WHAT'S OKAY?

It's normal not to be interested in sex at all. It's normal to like sex for purely physical reasons or for purely emotional reasons and to be either up for whatever or up only for doing, like, one specific thing that you've literally never heard of anyone else being up for. The brain is the biggest sexual organ, as they say, and it's not surprising that animals with noggins as complex as our own sometimes pine for complicated stuff in the sack. Like what you like. Don't stress about it.

And if you have sexual urges that drive you to hurt other people, you deserve compassionate help in managing those desires without causing harm. On the other hand, don't use kink as an excuse. "I know it's wrong to murder but I've just always really, really wanted to do it" would be a really shit murder defense, and "I know it's wrong to pull out my penis in front of my colleagues but I've just always really, really wanted to do it" is a really shit defense for dropping trou during a Zoom call. If you can't find a safe, sane, and consensual way to do what you want to do, don't do it. And if that means you're in anguish or you feel in danger of snapping, then you need help. There's nothing wrong with needing help. Look for "kink-aware" and "sex-positive" therapists to talk to, if you can't find someone who specializes in your specific paraphilic disorder.

To wrap up this chapter (and this book) on a lighter note, here are a few of my favorite historical perverts. Please recognize, my friends, that I have (mostly) only deigned to include individuals for which there is *solid* historical evidence of what most of us would consider kinky sexual preferences. Missing from this list are countless historical pervs whose sexual goings-on are mere hearsay—and truly, they are countless in number.

Catherine the Great, for example, is rumored in several corners of the internet to have employed full-time "foot ticklers" to help her indulge her kinks. Unfortunately, my deepest internet dive revealed several pages of porn-clip search results but no primary source for the historical "fact." Catherine the Great, of course, was the victim of several salacious rumors in and after her time (no, she did *not* fuck a *horse*), so we have to allow for the possibility that her tickle servants were similarly fabricated. You're welcome to entertain whatever niche headcanon you wish to assign to random historical figures you think might have been freaky. But I merely report the facts.

I've also left out an untold number of powerful historical men whose sexual weirdness is too wrapped up in abusive behavior to deserve inclusion. There's nothing fun or goofy about Casanova grooming children and buying sex slaves. In other words, there's a lot more ostensibly kinky stuff than this to be found in the historical canon. We're just here to have fun.

JAMES JOYCE, LOVER OF FARTS

This Irish novelist enjoyed a rambunctious sex life with his longtime partner and eventual wife Nora Barnacle, and we've got the letters to prove it. Indeed, after she unapologetically flaked on what was meant to be their first date in 1904, according to Joyce, she almost immediately gave him a hand job when they successfully met up. Thus began, if Joyce's letters to Barnacle are to be taken literally, a long and loving relationship that featured heaps of absolutely filthy sex.[14] According to Joyce, Barnacle awakened a real freak in him. He wrote as much to his "dirty little f*ckbird" in 1909, while away on a trip to get *Dubliners* published: "As you know, dearest, I never use obscene phrases in speaking. You have never heard me, have you, utter an unfit word before others. When men tell in my presence here filthy or lecherous stories I hardly smile. Yet you seem to turn me into a beast. It was you yourself, you naughty shameless girl who first led the way."

Later that year, he made it known just how much he valued even her farts as an erotic amusement:

You had an arse full of farts that night, darling, and I f*cked them out of you, big fat fellows, long windy ones, quick little merry cracks and a lot of tiny little naughty farties ending in a long gush from your hole. It is wonderful to f*ck a farting woman when every f*ck drives one out of her. I think I would know Nora's fart anywhere. I think I could pick hers out in a roomful of farting women. It is a rather girlish noise not like the wet windy fart which I imagine fat wives have. It is sudden and dry and dirty like what a bold girl would let off in fun in a school dormitory at night. I hope Nora will let off no end of her farts in my face so that I may know their smell also.

A flatulence fetish is otherwise known as eproctophilia. As Joyce's letters so colorfully demonstrate, it can be part of a perfectly lovely and devoted relationship.

Like all fetishes, eproctophilia can come in a wide variety of flavors and present in all sorts of ways. In one 2013 case study, a twenty-two-year-old with the fetish recalled that the fixation began when a grade-school crush let loose some gas in front of him, and while he was heterosexual in terms of his desires for "typical" intercourse, he had come to realize he got turned on by farts—the smellier, louder, and closer to his face the better—produced by people of any sex.[15]

BEN FRANKLIN, COUGAR HUNTER

While Ben Franklin ostensibly spent hours hanging out in the nude—a practice he called "air bathing"—for his health, he also made note of more overtly sexual practices he enjoyed. In a 1745 letter to a young friend advising on how best to deal with sexual urges before marrying, Franklin provided a thorough treatise in defense of banging older women in such circumstances.

Because he specifically mentions there being no risk of pesky conception as a reason in favor of aged mistresses, we can assume he's talking about aiming for the mid-forties or higher. Indeed, he advises his friend that wrinkles start in the face and take much longer to become apparent on clothed parts of the body, suggesting he might even mean significantly older. He makes several reasonable points (age begets experience in the bedroom and a better sense of discretion outside it) and several wholly offensive ones (as women become unattractive due to decrepitude, they focus more on learning how to physically and mentally please men as a substitute for prettiness, and they'll be more grateful for attention than young girls).

It's unlikely that Franklin had a particular fetish for elderly people—what's known as gerontophilia—because his other sexual exploits make it clear he was an equal-opportunity horndog. If anything, Franklin's extensive rationale for why certain kinds of women might make ideal targets for philandering gives him more in common with modern-day pickup artists than with a kinkster worthy of our admiration.

NERO, ANCIENT-DAY FURRY?

Yes, I promised we'd stick to reputable sources, but this one didn't exactly come from a supermarket tabloid. In 110 CE, the Roman historian Gaius Suetonius Tranquillus made some shocking claims about Emperor Nero, whose rule had ended some forty years previous. Among them was an anecdote about Nero dressing himself up in animal skins and being restrained like an animal, only so he could break loose and ravage all sorts of sexual partners in beast mode.[16]

Suetonius was basically a turn-of-the-millennium shock jockey, so we can't take anything he said seriously. But it's worth mentioning because it suggests that this practice existed somewhere in Roman society at the time, or else Suetonius had an absolutely prolific imagination.

Today, folks who like to dress up as animals and role-play corresponding personalities are generally known as furries, though the subculture features many people who don't get anything sexual out of the practice. Suetonius's little Nero fantasy also features what most modern kinksters would classify as "primal" sex—where one or both partners give in to and play up their more animalistic urges—and hints at (hopefully prenegotiated) consent play.

If physically overpowering someone (or being so overcome) sounds exciting to you, it's crucial that you communicate these plans and their implications with your partner and agree on limits and safe words beforehand. There's no shame in playing with the *idea* of consent, but don't put anyone's *actual* consent on the line to do so!

MOZART MAY HAVE BEEN PRO-RIMMING

I don't want to suggest that being into anilingus is particularly kinky. Frankly, putting your tongue on a part of a consenting partner's body is such an obvious way to interact sexually that making any of it taboo is absurd. Can butts be dirty? Yes. Can butts be clean? Also yes. The kinkiness of the act is, in my opinion, inversely correlated with the cleanliness of the butt.

Anyway, I don't know where else to talk about *Leck mich im Arsch* (Lick me in the ass), which is a canon sung in six-part round that Wolfgang Amadeus Mozart composed sometime around the 1780s. While it's often shared online as evidence that Wolfie liked a tongue in the rear, a closer look at the lyrics suggests it conveys more of a "you can kiss my ass" sentiment than a literal directive to french the singer's buttocks.

However, another song submitted by his widow for publication at the same time, *Leck mir den Arsch fein recht schön sauber* (Lick my ass nice and clean) seems a little less ambiguous, as he urges the listener to "come on, just try it," though it's now thought that Mozart merely wrote the lyrics to place them over the work of another composer.

JEAN-JACQUES ROUSSEAU, MISOGYNIST MASOCHIST

For all his writing on equality, Rousseau wasn't much of a feminist. His writing generally treated women with respect, but only when they were relegated to the sphere of domesticity and motherhood. In his private life, however, Rousseau seems to have enjoyed being on the receiving end of a powerful woman's disdain. In his midlife autobiography, published only after his death, he announced, "To be at the knees of an imperious mistress, to obey her orders, to have to beg her pardon, have been for me the sweetest delights."[17]

He traced this desire back to a childhood love of spanking, which was so overt that it led the female guardian who'd doled out the beatings to stop using them as a punishment.

There's an important juxtaposition here. While Rousseau wasn't a violent misogynist, at least according to historical reports, he also thought women should hold a submissive place in society. Why, then, did he want them to lord their sexual power over him in private?

Within kink communities, anecdotal evidence abounds of sexually submissive men with antifeminist politics or, more insidiously, a secret tendency to abuse and manipulate women. It's possible that the feeling something is wrong or difficult to obtain can add to sexual excitement for some people, opening the door for men who hate—or at least disrespect—women to enjoy getting a good walloping from one.

I point this out because, well, it might also help enlighten you as to why *you* have some kind of sexual fantasy that feels *totally* out of line with your actual life philosophy and also because it's important to realize that a kink does not an entire personality make. Don't let someone fool you into thinking they're harmless just because they want you to degrade them in the bedroom. A desperately submissive sexual partner can still prove capable of doling out physical and emotional abuse and can even wield their counterintuitive proclivities as a shield to hide the danger they pose.

MARY SHELLEY, GOTH GIRLFRIEND

I don't have any evidence that Mary Shelley, the genius who crafted *Frankenstein* while on an extremely moody summer holiday during a freak weather event, was kinky, per say. I just think you should know that she had sex for the first time on top of her mythical-status-feminist-icon mother's grave and also that, after her husband died, she's rumored to have kept his heart wrapped up in her desk. That may not be evidence of kinkiness, but it is, as the kids say, a whole mood.

Conclusion

EVERYTHING IS FINE
IF NOTHING HURTS

WHY WRITE A BOOK ABOUT THE ENTIRE HISTORY OF SEX? NO, REALLY, please tell me. Because at this moment—sitting alone in my apartment getting screamed at by a fifteen-year-old cat, desperately trying to shake off a sixty-hour workweek, and facing down a hard deadline for the manuscript—I'm having a tough time remembering.

But I think it goes a little something like this.

Once there was a girl growing up in the part of South Jersey that's basically Alabama. She knew more than most. Her mom was a gynecologist—one of the only female ones in that backroad county—and that meant some amount of sex-ed by osmosis. She was, for example, one of just a few members of her ninth-grade health class who weren't shocked to learn that HIV was sexually transmitted (yikes). She knew what menstruation was by the time she could talk (though she did assume that tampons worked like tub drain plugs until she was seven). Her first paying job was reorganizing a shelf of textbooks about vulvar disease at the age of seven (the great Blue Waffle saga of 2008 left her unscathed).

She grew up swiping donut-shaped pessary sample units to use as squishy stress toys, which meant she also grew up knowing that sometimes your uterus can make an appearance in your toilet bowl while you're just trying to take a shit.

But that wasn't the whole story.

She also went to an evangelical church. There, she learned to parrot fun catchphrases like "abortion is genocide" and listened to lectures about not wearing clothing that might "tempt her brothers in Christ." She started to realize this was all a bunch of unhelpful nonsense the day her youth pastor stood there, in his button-down polo and cargo shorts, and told her that holding hands with boys would demonstrably subtract from the amount of love she'd be able to feel for her future husband. But the messages she'd absorbed about intimacy didn't vanish as quickly as her admiration for Christian alt-rock.

Sex was something she knew she would have to keep secret. So, she did. And when it hurt, she assumed that was how it should be. She was taught that boys were animals who wanted nothing more than to deflower her; she was also taught that girls didn't much like sex at all. So, at sixteen, when that's how it was, she figured—well, yeah, that was how it was.

And then she grew up, and she thought she was empowered. She thought she'd learned all her lessons. But as years of therapy would one day teach her, humans are pretty good at repeating the same patterns over and over again in slightly different shades of gray.

She met a man who styled himself as free from all the societal baggage that weighed most people down around sex. He made her believe that the way to expand her horizons was to let him bulldoze through her boundaries. He hurt her a lot. For a long time. And then she decided to figure out who she was, *without treating sex like it carried the mortal weight of her soul.*

Here's what she learned: that man couldn't have been more wrong about what it meant to be sexually liberated. Because there's only one thing you can do in bed (or in a bathroom stall, or up against a wall, or whatever, that's your own damn business) that's actually messed up. The one bad thing you can do is cause harm to another living thing in the name of getting your rocks off. It's gauche. It's banal.

(And, like, hey, obviously sometimes you can inflict a little or a lot of *consensual* damage to someone, physically or psychologically. But the crucial thing there is that they want it *enthusiastically*, so that it's not something they're going to spend years talking about in therapy. Because while caring for your mental health is very sexy, being the reason that someone needs to get their head shrunk is decidedly not.)

What that girl learned about sex was key to her becoming a happy and healthy human (though one who still, as previously discussed, is scrambled by deadlines). This book shares what she learned.

1. Most of the rules we take for granted around sex and gender are arbitrary and harmful (which is especially messed up given how little work we put in, at the societal level, to ensure that people aren't using sex to hurt other people).
2. Sex doesn't have to be about making babies.

3. Being queer isn't new (not even a little), it isn't strange (though you can *be* as strange as you want to be), and it isn't nonsensical (no matter what anyone tells you).

4. It's okay not to be interested in sex—tonight, for a year, ever. That's perfectly normal. Wanting and having sex is not a prerequisite to a full and healthy adulthood and is not something you'll ever owe to anyone.

5. I promise that whatever it is you think is hot, you're not alone.

6. If you think you might be causing someone harm, stop. If what you want is inherently harmful to other living things, you can and should ask for help.

7. You're probably not masturbating more than a normal amount, though I can't promise you you're not an above-average student in that regard.

8. Pay attention to who holds the power. Are you sure you're comfortable? Are you sure your partner is comfortable? Or could one of you be feeling pressured?

9. We have always projected our cultural nonsense onto sex and how we have it and talk about it, which means it's always been complicated, and weird, and also the most normal thing in the world. What's normal now has not always been so and won't be forever.

10. I can't tell you what you need. But I can tell you that it's important to ask yourself what you want.

Here's the thing: you're just fine.

ACKNOWLEDGMENTS

MANY PEOPLE ARE TO BLAME FOR MY DUBIOUS DECISION TO WRITE THIS book. Because it's impossible to know where to begin, I'll start at the beginning: I am eternally grateful to my agent Jeff Shreve for buying me coffee, listening to me enthuse about herpes, and sticking with me through the two years it took to complete and sell my proposal (and even still). Thank you to Remy Cawley for giving this strange manuscript a home at Bold Type, and to my editor, Ben Platt, for convincing me that I could, in fact, write a book, and then convincing me over and over again.

Thank you to Sara Krolewski for her diligent fact-checking work, without which I'd never have had the guts to publish a single word. Thank you to Jen Kelland for the close and kind copyedit. Thank you to Katie Carruthers-Busser for ferrying this book through its production process, to Pete Garceau for giving it the cover art of my dreams, and to Hillary Brenhouse for overseeing the process from start to finish. Thank you to Johanna Dickson and Lindsay Fradkoff for their publicity and marketing expertise, and to all the friends and colleagues who gave this book a leg up in its early days.

Acknowledgments

Thank you to each and every person who asked me how this project was going, even when I really, *really* didn't want to tell you. Thank you especially to Paige, Josh, Lily, Arielle, Abe, and Ryan, who never got tired of hearing and somehow never get tired of me.

There were days when I didn't think I could actually manage writing a book on top of being the executive editor of a 150-year-old magazine. The fact that I could is due entirely to the grace and understanding and boundless support offered up by the *Popular Science* staff and extended Recurrent Media universe. Corinne, Purbita, Rob, John, Stan, Claire, Sara, Jess, Sara Kiley, Mike, Sandra, Charlotte, Jean, Katie, Russ, Chuck, Susan, Philip, Monroe, Ben, Lauren: Thank you.

My mom and dad are, naturally, entirely to blame for the person I am today, as are my beautiful and talented siblings. Arden and Chelsea: You have my whole heart, forever, and I am so glad the universe put us in the same time and place. Dad: Thank you for always being my number one fan and for putting my name on Mars, which I'm pretty sure means I own it. Mom: Thank you for showing me what strength looks like, and for never letting me forget that I am a woman of valor.

My heart is filled with gratitude for the therapists and doctors who helped me find the strength to heal when I was shattered. The abuse-survivor-to-sex-positive-nonfiction-book pipeline isn't exactly a popular waterway, so you clearly did something right.

To everyone I dated before those therapists got to me: Sorry, it was a weird time.

Last but not least, I owe this book and my life to the two most important people in it. Amy: To call you my best friend feels reductive. You are, more broadly, my best person, and I have more faith in the goodness of the world knowing that you exist in it.

To Oliver: Being stuck inside a strange new apartment in a strange new city during a strange new pandemic should have been worse than hell, even without balancing book deadlines with the crushing demands of a day job. Thank you for all the times you did laundry, ordered me noodles from Saigon Cafe, and accepted that I did *not* want to talk to you about how the book was going. Guess what: It's finished! Thanks for loving me enough to marry me twice. The feeling is mutual.

FURTHER READING

I'VE SAID IT BEFORE, BUT I'LL SAY IT AGAIN: WHO THE HECK AM I TO tell you what sex is? This book is meant to merely be a humorous primer, a lighthearted introduction, a romp through the basics. Here you'll find some suggestions for further reading.

Straight: The Surprisingly Short History of Heterosexuality by Hanne Blank (2012) did fantastic work of breaking my brain with its philosophizing on the nature of orientation. *The Invention of Heterosexuality* by Jonathan Ned Katz (1995) is another excellent read on the subject. For a much deeper dive into the history of love between women, I recommend *Sapphistries* by Leila J. Rupp (2009). For more detail on trans history, a subject in which my knowledge is woefully inadequate, try *Transgender History* by Susan Stryker (2017), *Black on Both Sides: A Racial History of Trans Identity* by C. Riley Snorton (2017), and *Female Husbands: A Trans History* by Jen Manion (2020).

I hope that readers on the Ace spectrum don't mistake my lack of pontificating on asexuality for a lack of regard. One of the most frustrating aspects of writing a book is that it can't be about everything, and I chose to focus more on the history of having sex than the lived experience of people who don't have it.

Luckily, Angela Chen's *Ace: What Asexuality Reveals About Desire, Society, and the Meaning of Sex* masterfully covers all of the bases I missed.

The idea of sex as a complex biological continuum may be new to some readers, but biologist Anne Fausto-Sterling presented it as a beneficial lens for scientists in *Sexing the Body: Gender Politics and the Construction of Sexuality* back in 2000. This book also tackles, in detail, the tragically common practice of inflicting genital mutilation on infants to "fix" intersex conditions. *Fixing Sex: Intersex, Medical Authority, and Lived Experience* by Katrina Karkazis (2009) is another invaluable book on these topics.

Labor of Love by Moira Weigel (2016) provides an interesting take on the institution of dating. *Virgin: The Untouched History* (2007) is another great read by Hanne Blank.

Buzz: The Stimulating History of the Sex Toy by Hallie Lieberman (2017), whose academic work I cite in breaking down misconceptions about the history of vibrators, is great for anyone looking for more info on self-pleasure. *A Curious History of Sex* by Kate Lister (2020) came out while my book was well underway but has a lot of the same DNA. Lister's *Whores of Yore* blog and Twitter account are incredible resources for anyone looking to learn more about sex.

Readers looking to do a deep (like, really deep) dive into the subject of baby making need look no further than *Reproduction: Antiquity to the Present Day* (2018), edited by Nick Hopwood, Rebecca Flemming, and Lauren Kassell. It's like the academic evil twin of the book you've just read (or perhaps it is I who am the more mischievous sibling here), and it's so chock-full of historical insight that I once almost shattered my glass coffee table with it. This also feels like as good a place as any to give my regards to *Bonk: The Curious Coupling of Science and Sex* (2008) by Mary Roach, which did not so much directly inform this particular book as it did directly inform the trajectory of my entire career.

Those craving more details on the history of sexual impotence and associated cures should check out *Impotence: A Cultural History* by Angus McLaren (1992). I also must credit the delightful documentary *Nuts!* by Penny Lane (2016) for helping kick off my research about John Romulus Brinkley, goat-testicle salesman extraordinaire.

While it didn't come out in time to inform the contents of my own manuscript, I wholeheartedly recommend *Hurts So Good: The Science and Culture of Pain on Purpose* by Leigh Cowart (2021). It's a downright necessary read for anyone with even the most casual interest in masochism (as a kink or just an abstract concept—you do you!). *Perv: The Sexual Deviant in All of Us* by Jesse Bering (2013) is another great book on the topic of kink.

NOTES

INTRODUCTION: EVERYTHING WEIRD IS NORMAL—EVERYTHING NORMAL IS WEIRD

1. Paul R. Ehrlich, David S. Dobkin, and Darryl Wheye, "Copulation," Stanford University, accessed October 10, 2021, https://web.stanford.edu/group/stanfordbirds/text/essays/Copulation.html.

2. Carl Zimmer, "The Sex Life of Birds, and Why It's Important," *New York Times*, June 6, 2013, www.nytimes.com/2013/06/06/science/the-sex-life-of-birds-and-why-its-important.html.

3. Jason G. Goldman, "Duck Penises Grow Bigger Among Rivals," *National Geographic*, May 3, 2021, www.nationalgeographic.com/animals/article/duck-penis-size-social-group-study.

4. Pierre Mineau, Frank McKinney, and Scott R. Derrickson, "Forced Copulation in Waterfowl," *Behaviour* 86, nos. 3–4 (1983): 250–293, https://doi.org/10.1163/156853983x00390; Carl Zimmer, "In Ducks, War of the Sexes Plays Out in the Evolution of Genitalia," *New York Times*, May 1, 2007, www.nytimes.com/2007/05/01/science/01duck.html.

5. Matt Kaplan, "The Sex Wars of Ducks," *Nature*, 2009, https://doi.org/10.1038/news.2009.1159.

6. Patricia L. R. Brennan et al., "Coevolution of Male and Female Genital Morphology in Waterfowl," *PLoS ONE* 2, no. 5 (2007), https://doi.org/10.1371/journal.pone.0000418.

1. WHAT THE HECK IS SEX?

1. S. Otto, "Sexual Reproduction and the Evolution of Sex," *Nature Education* 1, no. 1 (2008): 182.

2. M. A. O'Malley et al., "Concepts of the Last Eukaryotic Common Ancestor," *Nature Ecology & Evolution* 3 (2019): 338–344, https://doi.org/10.1038/s41559-019-0796-3.

3. Michelle Starr, "Your Body Makes 3.8 Million Cells Every Second. Most of Them Are Blood," ScienceAlert, January 23, 2021, www.sciencealert.com/your-body-makes-4-million-cells-a-second-and-most-of-them-are-blood.

4. Christie Wilcox, "Researchers Rethink the Ancestry of Complex Cells," *Quanta*, April 9, 2019, www.quantamagazine.org/rethinking-the-ancestry-of-the-eukaryotes-20190409.

5. Nicholas J. Butterfield, "Bangiomorpha Pubescensn. Gen., n. Sp.: Implications for the Evolution of Sex, Multicellularity, and the Mesoproterozoic/Neoproterozoic Radiation of Eukaryotes," *Paleobiology* 26, no. 3 (2000): 386–404, https://doi.org/10.1666/0094-8373(2000)026<0386:bpngns>2.0.co;2.

6. "Origins of Sex Discovered: Side-by-Side Copulation in Distant Ancestors," ScienceDaily, October 20, 2014, https://www.sciencedaily.com/releases/2014/10/141020103840.htm.

7. "Flinders Scientist Discovers Origins of Sex," *Flindersblogs*, October 19, 2014, https://news.flinders.edu.au/blog/2014/10/20/flinders-scientist-discovers-origins-of-sex.

8. A. S. Kondrashov, "Classification of Hypotheses on the Advantage of Amphimixis," *Journal of Heredity* 84, no. 5 (1993): 372–387, https://pubmed.ncbi.nlm.nih.gov/8409359.

9. Abigail J. Lynch, "Why Is Genetic Diversity Important?," US Geological Survey, April 26, 2016, www.usgs.gov/center-news/why-genetic-diversity-important.

10. Randy C. Ploetz, "Panama Disease: A Classic and Destructive Disease of Banana," *Plant Health Progress* 1, no. 1 (December 4, 2000): 10, https://doi.org/10.1094/php-2000-1204-01-hm.

11. "Research Affirms Sexual Reproduction Avoids Harmful Mutations," Phys.org, January 12, 2015, https://phys.org/news/2015-01-affirms-sexual-reproduction-mutations.html.

12. Nina Gerber et al., "Daphnia Invest in Sexual Reproduction When Its Relative Costs Are Reduced," *Proceedings of the Royal Society B: Biological Sciences* 285, no. 1871 (2018): 20172176, https://doi.org/10.1098/rspb.2017.2176.

13. Alan W. Gemmill, Mark E. Viney, and Andrew F. Read, "Host Immune Status Determines Sexuality in a Parasitic Nematode," *Evolution* 51, no. 2 (1997): 393, https://doi.org/10.2307/2411111.

14. Rachel Feltman, "A Female Shark Had a Bunch of Babies Without Male Contact," *Popular Science*, January 18, 2018, www.popsci.com /parthenogenesis-shark-reproduction-without-males.

15. Rachel Feltman, "Scientists Examine Why Men Even Exist," *Washington Post*, May 18, 2015, www.washingtonpost.com/news/speaking-of-science /wp/2015/05/18/scientists-examine-why-men-even-exist.

2. HOW NORMAL IS HETERONORMATIVITY?

1. M. Scott Taylor, "Buffalo Hunt: International Trade and the Virtual Extinction of the North American Bison," National Bureau of Economic Research, March 2007, https://doi.org/10.3386/w12969.

2. H. Vervaecke and C. Roden, "Going with the Herd: Same-Sex Interaction and Competition in American Bison," in *Homosexual Behaviour in Animals: An Evolutionary Perspective*, ed. V. Sommer. and P. Vasey, 131–153 (Cambridge: Cambridge University Press, 2006), www.researchgate.net /publication/272832692_Vervaecke_H_Roden_C_2006_Going_with_the _herd_same-sex_interaction_and_competition_in_American_bison_In _Homosexual_Behaviour_in_Animals_An_Evolutionary_Perspective _Ed_V_Sommer_P_Vasey_Cambridge_Univers.

3. Jana Bommersbach, "Homos on the Range," *True West Magazine*, November 1, 2005, https://truewestmagazine.com/old-west-homosexuality -homos-on-the-range.

4. Trudy Ring, "Merriam-Webster Updates Definition of 'Bisexual,'" *Advocate*, September 24, 2020, www.advocate.com/media/2020/9/23 /merriam-webster-updates-definition-bisexual.

5. Grace Wade, "Why Do Some Animals Engage in Same-Sex Sexual Behavior?," *Popular Science*, December 2, 2019, www.popsci.com/story/animals /same-sex-sexual-behavior-evolution.

6. Jeremy Yoder, "The Intelligent Homosexual's Guide to Natural Selection and Evolution, with a Key to Many Complicating Factors," *Scientific American*, June 21, 2011, https://blogs.scientificamerican.com/guest-blog /the-intelligent-homosexuals-guide-to-natural-selection-and-evolution -with-a-key-to-many-complicating-factors.

7. Mwenza Blell, "Grandmother Hypothesis, Grandmother Effect, and Residence Patterns," *International Encyclopedia of Anthropology*, 2017, 1–5, https://doi.org/10.1002/9781118924396.wbiea2162.

8. Paul L. Vasey and Doug P. VanderLaan, "An Adaptive Cognitive Dissociation Between Willingness to Help Kin and Nonkin in

Samoan Fa'afafine," *Psychological Science* 21, no. 2 (2010): 292–297, https://doi
.org/10.1177/0956797609359623.

9. Andrew B. Barron and Brian Hare, "Prosociality and a Sociosexual Hypothesis for the Evolution of Same-Sex Attraction in Humans," *Frontiers in Psychology* 10 (2020), https://doi.org/10.3389/fpsyg.2019.02955.

10. Robin McKie, "'Sexual Depravity' of Penguins That Antarctic Scientist Dared Not Reveal," *Guardian*, June 9, 2012, www.theguardian.com
/world/2012/jun/09/sex-depravity-penguins-scott-antarctic.

11. Dr. George Murray Levick, "Unpublished Notes on the Sexual Habits of the Adélie Penguin," Research Gate, October 2012,
www.researchgate.net/publication/259425517_Dr_George_Murray
_Levick_1876-1956_Unpublished_notes_on_the_sexual_habits_of_the
_Adelie_penguin.

12. Dinitia Smith, "Love That Dare Not Squeak Its Name," *New York Times*, February 7, 2004, www.nytimes.com/2004/02/07/arts/love-that-dare
-not-squeak-its-name.html.

13. "Cockchafer," Wikipedia, https://en.wikipedia.org/wiki/Cockchafer.

14. Marco Riccucci, "Same-Sex Sexual Behaviour in Bats," *Hystrix, the Italian Journal of Mammalogy*, 2011, https://doi.org/10.4404/Hystrix-22
.1-4478.

15. V. Wai-Ping and M. B. Fenton, "Nonselective Mating in Little Brown Bats (*Myotis lucifugus*)," *Journal of Mammalogy* 69, no. 3 (1988): 641–645,
https://doi.org/10.2307/1381364.

16. L. W. Braithwaite, "Ecological Studies of the Black Swan III. Behaviour and Social Organisation," *Wildlife Research* 8 (1981): 135–146, https://
doi.org/10.1071/WR9810135.

17. Yvette Tan, "New Zealand Goose: How One Blind Bisexual Bird Became an Icon," BBC News, February 17, 2018, www.bbc.com/news
/world-asia-43054363.

18. David M. Halperin, "Is There a History of Sexuality?," *History and Theory* 28, no. 3 (1989): 257, https://doi.org/10.2307/2505179.

19. Leila J. Rupp, "Toward a Global History of Same-Sex Sexuality," *Journal of the History of Sexuality* 10, no. 2 (April 2001): 287–302, www.jstor.org
/stable/3704817.

20. Catherine S. Donnay, "Pederasty in Ancient Greece: A View of a Now Forbidden Institution," *EWU Masters Thesis Collection* 506, EWU Digital Commons, 2018, https://dc.ewu.edu/theses/506.

21. "Symposium by Plato," Internet Classics Archive, accessed October 10, 2021, http://classics.mit.edu/Plato/symposium.html.

22. Sarah Prager, "In Han Dynasty China, Bisexuality Was the Norm," JSTOR Daily, June 10, 2020, https://daily.jstor.org/in-han-dynasty-china -bisexuality-was-the-norm.

23. Steven Dryden, "A Short History of LGBT Rights in the UK," British Library, www.bl.uk/lgbtq-histories/articles/a-short-history-of -lgbt-rights-in-the-uk.

24. Anne Lister, www.annelister.co.uk; "Anne Lister and Shibden Hall," Historic England, https://historicengland.org.uk/research/inclusive -heritage/lgbtq-heritage-project/love-and-intimacy/anne-lister-and-shibden -hall; "Same-Sex Marriages," Historic England, https://historicengland.org .uk/research/inclusive-heritage/lgbtq-heritage-project/love-and-intimacy /same-sex-marriages.

25. Sean Coughlan, "The 200-Year-Old Diary That's Rewriting Gay His-tory," BBC News, February 10, 2020, www.bbc.com/news/education-51385884.

26. "Boston Marriages," National Park Service, www.nps.gov/articles/000 /boston-marriages.htm.

27. Kasey Edwards, "A Boston Marriage: When You're More Than Friends but Less Than Lovers," *Sydney Morning Herald*, February 25, 2020, www .smh.com.au/lifestyle/life-and-relationships/a-boston-marriage-when-you-re -more-than-friends-but-less-than-lovers-20200224-p543ri.html.

28. Alex Ross, "Berlin Story," *New Yorker*, January 19, 2015, www.new yorker.com/magazine/2015/01/26/berlin-story.

29. "The First Institute for Sexual Science (1919–1933)," Magnus-Hirschfeld-Gesellschaft e.V., https://magnus-hirschfeld.de/ausstellungen /institute.

30. John Broich, "How the Nazis Destroyed the First Gay Rights Movement," Conversation, July 5, 2017, https://theconversation.com /how-the-nazis-destroyed-the-first-gay-rights-movement-80354.

31. Samantha Schmidt, "1 in 6 Gen Z Adults Are LGBT. And This Num-ber Could Continue to Grow," *Washington Post*, February 24, 2021, www .washingtonpost.com/dc-md-va/2021/02/24/gen-z-lgbt.

32. Samantha Allen, "Millennials Are the Gayest Genera-tion," Daily Beast, updated April 14, 2017, www.thedailybeast.com /millennials-are-the-gayest-generation.

3. JUST HOW MANY SEXES ARE THERE?

1. Logan D. Dodd et al., "Active Feminization of the Preoptic Area Occurs Independently of the Gonads in Amphiprion Ocellaris," *Hormones and Behav-ior* 112 (2019): 65–76, https://doi.org/10.1016/j.yhbeh.2019.04.002.

2. Rachel Feltman, "What Has Hundreds of Sexes and Excels at Math? This Is Slime Molds 101," *Popular Science*, October 17, 2019, www.popsci.com /the-blob-slime-mold.

3. Erika Kothe, "Mating Types and Pheromone Recognition in the Homobasidiomycete Schizophyllum Commune," *Fungal Genetics and Biology* 27, nos. 2–3 (1999): 146–152, https://doi.org/10.1006/fgbi.1999.1129.

4. Jacqueline Jacob and F. Ben Mather, "Sex Reversal in Chickens," University of Florida Extension, accessed October 10, 2021, https://ufdcimages.uflib .ufl.edu/IR/00/00/30/37/00001/PS05000.pdf.

5. "Episode 018: Spontaneous Sex Reversal in Chickens—My Hen Just Became a Rooster!," *Urban Chicken Podcast*, www.urbanchickenpodcast.com /ucp-episode-018.

6. "Hyena," Medieval Bestiary, http://bestiary.ca/beasts/beast153.htm.

7. Stephen Jay Gould, "Hyena Myths and Realities," University of British Columbia, Department of Zoology, www.zoology.ubc.ca/~bio336/Bio33 6/Readings/GouldHyena1981.pdf.

8. Holger Funk, "R. J. Gordon's Discovery of the Spotted Hyena's Extraordinary Genitalia in 1777," *Journal of the History of Biology* 45, no. 2 (summer 2012): 301–328, www.jstor.org/stable/41488454?seq=1.

9. Marion L. East, Heribert Hofer, and Wolfgang Wickler, "The Erect 'Penis' Is a Flag of Submission in a Female-Dominated Society: Greetings in Serengeti Spotted Hyenas," *Behavioral Ecology and Sociobiology* 33, no. 6 (1993), https://doi.org/10.1007/bf00170251.

10. Carrie Arnold, "The Sparrow with Four Sexes," *Nature* 539 (2016): 482–484, www.nature.com/articles/539482a.

11. M. Elaina et al., "Divergence and Functional Degradation of a Sex Chromosome–Like Supergene," *Current Biology* 26, no. 3 (2016): 344–350, https://doi.org/10.1016/j.cub.2015.11.069.

12. N. Uma Maheswari et al., "'Early Baby Teeth': Folklore and Facts," *Journal of Pharmacy and Bioallied Sciences* 4, no. 6 (2012): 329, https://doi .org/10.4103/0975-7406.100289.

13. Melanie Blackless et al., "How Sexually Dimorphic Are We? Review and Synthesis," *American Journal of Human Biology* 12, no. 2 (2000): 151–166, https://doi.org/10.1002/(sici)1520-6300(200003/04)12:2<151:aid-ajhb1>3.0 .co;2-f.

14. Mary García-Acero et al., "Disorders of Sexual Development: Current Status and Progress in the Diagnostic Approach," *Current Urology* 13, no. 4 (2020): 169–178, https://doi.org/10.1159/000499274.

15. Minghao Liu, Swetha Murthi, and Leonid Poretsky, "Polycystic Ovary Syndrome and Gender Identity," *Yale Journal of Biology and Medicine* 93, no. 4 (2020): 529–537, www.ncbi.nlm.nih.gov/pmc/articles/PMC7513432.

16. Chandra S. Pundir et al., "The Prevalence of Polycystic Ovary Syndrome: A Brief Systematic Review," *Journal of Human Reproductive Sciences* 13, no. 4 (2020): 261, https://doi.org/10.4103/jhrs.jhrs_95_18.

17. Judson J. Van Wyk and Ali S. Calikoglu, "Should Boys with Micropenis Be Reared as Girls?," *Journal of Pediatrics* 134, no. 5 (1999): 537–538, https://doi.org/10.1016/S0022-3476(99)70236-2.

18. Karen Lin-Su and Maria I. New, "Ambiguous Genitalia in the Newborn," *Avery's Diseases of the Newborn*, 2012, 1286–1306, https://doi.org/10.1016/b978-1-4377-0134-0.10092-7.

19. John Colapinto, *As Nature Made Him: The Boy Who Was Raised a Girl* (New York: HarperCollins, 2006).

20. John Colapinto, "Why Did David Reimer Commit Suicide?," *Slate*, June 3, 2004, https://slate.com/technology/2004/06/why-did-david-reimer-commit-suicide.html.

21. John Colapinto, "The True Story of John/Joan," *Rolling Stone*, December 11, 1992, https://web.archive.org/web/20000815095602/http:/www.pfc.org.uk/news/1998/johnjoan.htm.

22. Associated Press, "David Reimer, 38, Subject of the John/Joan Case (Published 2004)," *New York Times*, May 12, 2004, www.nytimes.com/2004/05/12/us/david-reimer-38-subject-of-the-john-joan-case.html.

23. "'I Want to Be like Nature Made Me,'" Human Rights Watch, December 15, 2020, www.hrw.org/report/2017/07/25/i-want-be-nature-made-me/medically-unnecessary-surgeries-intersex-children-us.

24. Emilienne Malfatto and Jelena Prtoric, "Last of the Burrnesha: Balkan Women Who Pledged Celibacy to Live as Men," *Guardian*, August 5, 2014, www.theguardian.com/world/2014/aug/05/women-celibacy-oath-men-rights-albania.

25. "The Third Gender and Hijras: Hinduism Case Study, 2018," Harvard Divinity School: Religion and Public Life, https://rpl.hds.harvard.edu/religion-context/case-studies/gender/third-gender-and-hijras.

26. Jon Letman, "'Mahu' Demonstrate Hawaii's Shifting Attitudes Toward LGBT Life," *Al Jazeera*, January 9, 2016, http://america.aljazeera.com/articles/2016/1/9/mahu-hawaii-gender-LGBT-acceptance.html.

27. Alan Weedon, "Fa'afafine, Fakaleitī, Fakafifine—Understanding the Pacific's Alternative Gender Expressions," *ABC News*, August 30, 2019,

www.abc.net.au/news/2019-08-31/understanding-the-pacifics-alternative
-gender-expressions/11438770.

28. "Inqueery: Indigenous Identity and the Significance of the Term
'Two-Spirit,'" *Them*, December 12, 2018, www.them.us/story/inqueery-two
-spirit.

29. "Native Identity and Tribal Sovereignty," Funders for LGBTQ Issues,
www.lgbtracialequity.org/perspectives/perspective.cfm?id=20.

30. J. P. Bogart et al., "Sex in Unisexual Salamanders: Discovery of a New
Sperm Donor with Ancient Affinities," *Heredity* 103, no. 6 (2009): 483–493,
https://doi.org/10.1038/hdy.2009.83.

31. J. P. Bogart, "Unisexual Salamanders in the Genus Ambystoma,"
Herpetologica 75, no. 4 (2019): 259, https://doi.org/10.1655/herpetologica
-d-19-00043.1.

4. HOW DO WE DO IT?

1. Justin R. Garcia et al., "Sexual Hook-Up Culture: A Review," *Review of
General Psychology* 16, no 2 (2012): 161–176, www.apa.org/monitor/2013/02
/ce-corner.

2. Marten Stol, *Women in the Ancient Near East* (Berlin: De Gruyter,
2016).

3. W. Haak et al., "Ancient DNA, Strontium Isotopes, and Osteological
Analyses Shed Light on Social and Kinship Organization of the Later Stone
Age," *Proceedings of the National Academy of Sciences* 105, no. 47 (2008): 18226–
18231, https://doi.org/10.1073/pnas.0807592105.

4. M. Dyble et al., "Sex Equality Can Explain the Unique Social Structure
of Hunter-Gatherer Bands," *Science* 348, no. 6236 (2015): 796–798, https://
doi.org/10.1126/science.aaa5139.

5. S. Beckerman and P. Valentine, "Introduction: The Concept of Partible
Paternity Among Native South Americans," in *Cultures of Multiple Fathers: The
Theory and Practice of Partible Paternity in Lowland South America*, ed. S. Beckerman and P. Valentine, 1–13 (Gainesville: University Press of Florida, 2002),
http://radicalanthropologygroup.org/sites/default/files/pdf/class_text_050
.pdf.

6. Robert S. Walker, Mark V. Flinn, and Kim Hill, "Evolutionary History
of Partible Paternity in South America," *Proceedings of the National Academy
of Sciences of the United States of America* 107, no. 45 (October 2010): 19195–
19200, www.researchgate.net/publication/47544385_Evolutionary_History
_of_Partible_Paternity_in_South_America.

7. Alexandra Genova, "Where Women Reign: An Intimate Look Inside a Rare Kingdom," *National Geographic*, August 14, 2017, www.nationalgeographic.com/photography/article/portraits-of-chinese-Mosuo-matriarchs.

8. For more on polyandry, see Laura A. Benedict, "Polyandry Around the World," Digital Scholarship@UNLV, 2017, https://digitalscholarship.unlv.edu/cgi/viewcontent.cgi?article=1139&context=award.

9. Julian H. Steward, "Shoshoni Polyandry," *American Anthropologist* 38, no. 4 (1936): 561–564, https://doi.org/10.1525/aa.1936.38.4.02a00050.

10. Martin Surbeck et al., "Male Reproductive Skew Is Higher in Bonobos Than Chimpanzees," *Current Biology* 27, no. 13 (2017), https://doi.org/10.1016/j.cub.2017.05.039.

11. Martin Surbeck et al., "Males with a Mother Living in Their Group Have Higher Paternity Success in Bonobos but Not Chimpanzees," *Current Biology* 29, no. 10 (2019), https://doi.org/10.1016/j.cub.2019.03.040.

12. Paul Cartledge, "Spartan Wives: Liberation or Licence?," *Classical Quarterly, New Series* 31, no. 1 (1981): 84–105, https://faculty.uml.edu//ethan_spanier/teaching/documents/cartledgespartanwomen.pdf.

13. M. C. Andrade, "Risky Mate Search and Male Self-Sacrifice in Redback Spiders," *Behavioral Ecology* 14, no. 4 (2003): 531–538, https://doi.org/10.1093/beheco/arg015.

14. Matthias W. Foellmer and Daphne J. Fairbairn, "Spontaneous Male Death During Copulation in an Orb-Weaving Spider," *Proceedings of the Royal Society of London. Series B: Biological Sciences* 270, suppl. 2 (2003), https://doi.org/10.1098/rsbl.2003.0042.

15. Jeremy B. Swann et al., "The Immunogenetics of Sexual Parasitism," *Science* 369, no. 6511 (2020): 1608–1615, www.science.org/lookup/doi/10.1126/science.aaz9445.

16. Katherine J. Wu, "How the Ultimate Live-In Boyfriend Evolved His Way Around Rejection," *New York Times*, July 30, 2020, www.nytimes.com/2020/07/30/science/anglerfish-immune-rejection.html.

17. William Keener et al., "The Sex Life of Harbor Porpoises (*Phocoena phocoena*): Lateralized and Aerial Behavior," *Aquatic Mammals* 44 (2018): 620–632, www.researchgate.net/publication/328966249_The_Sex_Life_of_Harbor_Porpoises_Phocoena_phocoena_Lateralized_and_Aerial_Behavior.

18. Laura Marjorie Miller, "The Biomechanics of Sex," *UMASS*, Summer 2017, www.umass.edu/magazine/summer-2017/biomechanics-sex.

19. Eldon Greij, "Manakins' Wild Courtship Rituals Explained," *Bird-Watching*, October 4, 2018, www.birdwatchingdaily.com/news/science/manakins-wild-courtship-rituals-explained.

20. "Pheromones in Insects," Smithsonian, www.si.edu/spotlight/buginfo/pheromones.

21. D. Trotier, "Vomeronasal Organ and Human Pheromones," *European Annals of Otorhinolaryngology, Head and Neck Diseases* 128, no. 4 (2011): 184–190, https://doi.org/10.1016/j.anorl.2010.11.008.

22. Agata Groyecka et al., "Attractiveness Is Multimodal: Beauty Is Also in the Nose and Ear of the Beholder," *Frontiers in Psychology* 8 (May 2017), www.frontiersin.org/article/10.3389/fpsyg.2017.00778.

23. Bob Yirka, "Evidence That Humans Prefer Genetically Dissimilar Partners Based on Scent," Phys.org, March 20, 2019, https://phys.org/news/2019-03-evidence-humans-genetically-dissimilar-partners.html.

24. Samantha Joel et al., "Machine Learning Uncovers the Most Robust Self-Report Predictors of Relationship Quality Across 43 Longitudinal Couples Studies," *Proceedings of the National Academy of Sciences* 117, no. 32 (2020): 19061–19071, https://doi.org/10.1073/pnas.1917036117.

25. Erin Blakemore, "What's the Secret to Sexiness?," *Popular Science*, September 15, 2021, www.popsci.com/science/human-attraction-hotness.

26. Natalie Zarrelli, "The Awkward 17th-Century Dating Practice That Saw Teens Get Bundled into Bags," Atlas Obscura, January 26, 2017, www.atlasobscura.com/articles/the-awkward-17thcentury-dating-practice-that-saw-teens-get-bundled-into-bags.

27. "The Victorian Craze That Sparked a Mini–Sexual Revolution," BBC News Magazine, April 6, 2015, www.bbc.com/news/magazine-31831110.

28. Holly Furneaux, "Victorian Sexualities," British Library, May 15, 2014, www.bl.uk/romantics-and-victorians/articles/victorian-sexualities.

29. "Cherry Picking: A History of Testing Virginity," *Whores of Yore*, August 21, 2017, www.thewhoresofyore.com/katersquos-journal/cherry-picking-a-history-of-testing-virginity.

30. Moira Weigel, *Labor of Love: The Invention of Dating* (New York: Farrar, Straus and Giroux, 2017).

31. Linton Weeks, "When 'Petting Parties' Scandalized the Nation," NPR, May 26, 2015, www.npr.org/sections/npr-history-dept/2015/05/26/409126557/when-petting-parties-scandalized-the-nation.

32. "Infographic: A History of Love and Technology," POV, http://archive.pov.org/xoxosms/infographic-technology-dating.

33. "'Bohemian, Broad-Minded, Unconventional.' What Was It like to Be Queer in the 1920s?," National Archives, November 15, 2019, https://blog.nationalarchives.gov.uk/bohemian-broad-minded-unconventional-what-was-it-like-to-be-queer-in-the-1920s.

34. Christine Foster, "Punch-Card Love," *Stanford Magazine*, March 2007, https://stanfordmag.org/contents/punch-card-love; Alicia M. Chen, "Operation Match," *Harvard Crimson*, February 16, 2018, www.thecrimson.com/article/2018/2/16/operation-match.

35. Michael Rosenfeld, Reuben J. Thomas, and Sonia Hausen, "Disintermediating Your Friends: How Online Dating in the United States Displaces Other Ways of Meeting," *Proceedings of the National Academy of Sciences of the United States of America* 116, no. 36 (July 15, 2019), https://doi.org/10.1073/pnas.1908630116.

36. Eli J. Finkel et al., "Online Dating," *Psychological Science in the Public Interest* 13, no. 1 (2012): 3–66, https://doi.org/10.1177/1529100612436522.

37. Maddie Holden, "Gen Z Are 'Puriteens,' but Not for the Reasons You Think," *GQ*, July 30, 2021, www.gq.com/story/gen-z-puriteens.

5. WHAT'S THE DEAL WITH MASTURBATION?

1. Sarah Laskow, "Everything You've Heard About Chastity Belts Is a Lie," Atlas Obscura, July 12, 2017, www.atlasobscura.com/articles/everything-youve-heard-about-chastity-belts-is-a-lie.

2. "Genesis," Bonobo Sexuality and Behavior, Reed College, www.reed.edu/biology/courses/BIO342/2011_syllabus/2011_websites/subramanian_jaime/genesis.html.

3. Ben G. Blount, "Issues in Bonobo (*Pan Paniscus*) Sexual Behavior," *American Anthropologist* 92, no. 3 (1990): 702–714, https://doi.org/10.1525/aa.1990.92.3.02a00100.

4. Randall L. Susman, *The Pygmy Chimpanzee: Evolutionary Biology and Behavior* (New York: Plenum Press, 1984).

5. "Pre-copulatory Ejaculation Solves Time Constraints During Copulations in Marine Iguanas," *Proceedings of the Royal Society of London. Series B: Biological Sciences* 263, no. 1369 (1996): 439–444, https://doi.org/10.1098/rspb.1996.0066.

6. N. Gunst, P. L. Vasey, and J. B. Leca, "Deer Mates: A Quantitative Study of Heterospecific Sexual Behaviors Performed by Japanese Macaques Toward Sika Deer," *Archives of Sexual Behavior* 47 (2018): 847–856, https://link.springer.com/article/10.1007/s10508-017-1129-8.

7. Christian Tighe, "Masturbation Has Evolved for the Better," *Vice*, July 17, 2017, www.vice.com/en/article/evdm8m/masturbation-has-evolved-for-the-better.

8. Yao-Hua Law, "Masturbating Macaques Give Scientists a Hand with Semen Collection," Earth Touch News Network, May 26, 2014, www.earthtouchnews.com/natural-world/animal-behaviour/masturbating-macaques-give-scientists-a-hand-with-semen-collection.

9. Vern L. Bullough, "Masturbation," *Journal of Psychology & Human Sexuality* 14, nos. 2–3 (2003): 17–33, www.tandfonline.com/doi/abs/10.1300/J056v14n02_03.

10. Bullough, "Masturbation."

11. A. Walthall, "Masturbation and Discourse on Female Sexual Practices in Early Modern Japan," *Gender & History* 21 (2009): 1–18, https://onlinelibrary.wiley.com/doi/abs/10.1111/j.1468-0424.2009.01532.x.

12. Paige Donaghy, "Lascivious Virgins and Lustful Itches: Women's Masturbation in Early England," Conversation, March 24, 2019, https://theconversation.com/lascivious-virgins-and-lustful-itches-womens-masturbation-in-early-england-101260.

13. "Masculinity, Pornography, and the History of Masturbation," *Sexuality & Culture* 16 (2012): 306–320, https://link.springer.com/article/10.1007/s12119-011-9125-y; "Onanism, or, a Treatise upon the Disorders Produced by Masturbation: Or, the Dangerous Effects of Secret and Excessive Venery / by M. Tissot; Translated from the Last Paris Edition, by A. Hume," Wellcome Collection, accessed October 10, 2021, https://wellcomecollection.org/works/y7fcmq69/items?canvas=57.

14. J. Mortimer Granville, "Nerve-Vibration and Excitation as Agents in the Treatment of Functional Disorder and Organic Disease [1883]," HathiTrust, accessed October 10, 2021, https://babel.hathitrust.org/cgi/pt?id=nnc2.ark%3A%2F13960%2Ft29890j4d&view=1up&seq=61&skin=2021.

15. Hallie Lieberman and Eric Schatzberg, "A Failure of Academic Quality Control: The Technology of Orgasm," *Journal of Positive Sexuality* 4, no. 2 (2018): 24–47, https://doi.org/10.51681/1.421.

16. Ashley Fetters and Robinson Meyer, "Victorian-Era Orgasms and the Crisis of Peer Review," *Atlantic*, September 7, 2018, www.theatlantic.com/health/archive/2018/09/victorian-vibrators-orgasms-doctors/569446.

17. Hallie Lieberman, "(Almost) Everything You Know About the Invention of the Vibrator Is Wrong," *New York Times*, January 23, 2020, www.nytimes.com/2020/01/23/opinion/vibrator-invention-myth.html.

18. Adee Braun, "Looking to Quell Sexual Urges? Consider the Graham Cracker," *Atlantic*, January 15, 2014, www.theatlantic.com/health /archive/2014/01/looking-to-quell-sexual-urges-consider-the-graham -cracker/282769.

19. Kyla Wazana Tompkins, "Sylvester Graham's Imperial Dietetics," *Gastronomica* 9, no. 1 (2009): 50–60, https://doi.org/10.1525/gfc.2009.9.1.50.

20. Danny Lewis, "American Vegetarianism Has a Religious Past," *Smithsonian Magazine*, August 20, 2015, www.smithsonianmag.com/smart-news /american-vegetarianism-had-religious-upbringing-180956346.

21. Dan MacGuill, "Were Kellogg's Corn Flakes Created as an 'Antimasturbatory Morning Meal'?," Snopes.com, accessed October 11, 2021, www.snopes.com/fact-check/kelloggs-corn-flakes-masturbation.

22. Howard Markel, "The Secret Ingredient in Kellogg's Corn Flakes Is Seventh-day Adventism," *Smithsonian Magazine*, July 28, 2017, www .smithsonianmag.com/history/secret-ingredient-kelloggs-corn-flakes -seventh-day-adventism-180964247.

6. WHY ARE WE SO SCARED OF STIS?

1. Katherine E. Dahlhausen et al., "Characterization of Shifts of Koala (*Phascolarctos Cinereus*) Intestinal Microbial Communities Associated with Antibiotic Treatment," *PeerJ* 6 (2018), https://doi.org/10.7717/peerj.4452.

2. Amy Robbins et al., "Longitudinal Study of Wild Koalas (*Phascolarctos Cinereus*) Reveals Chlamydial Disease Progression in Two Thirds of Infected Animals," *Scientific Reports* 9, no. 1 (2019), https://doi.org/10.1038 /s41598-019-49382-9.

3. Ville Pimenoff et al., "The Role of Adna in Understanding the Coevolutionary Patterns of Human Sexually Transmitted Infections," *Genes* 9, no. 7 (2018): 317, https://doi.org/10.3390/genes9070317.

4. R. Eberle and L. Jones-Engel, "Understanding Primate Herpesviruses," *Journal of Emerging Diseases and Virology* 3, no. 1 (2017), https://doi .org/10.16966/2473-1846.127.

5. Simon J. Underdown, Krishna Kumar, and Charlotte Houldcroft, "Network Analysis of the Hominin Origin of Herpes Simplex Virus 2 from Fossil Data," *Virus Evolution* 3, no. 2 (2017), https://doi.org/10.1093/ve/vex026.

6. Houssein H. Ayoub, Hiam Chemaitelly, and Laith J. Abu-Raddad, "Characterizing the Transitioning Epidemiology of Herpes Simplex Virus Type 1 in the USA: Model-Based Predictions," *BMC Medicine* 17, no. 1 (2019), https://doi.org/10.1186/s12916-019-1285-x.

7. Jason Daley, "Neanderthals May Have Given Us Both Good Genes and Nasty Diseases," *Smithsonian Magazine*, October 20, 2016, www .smithsonianmag.com/smart-news/new-studies-show-neanderthals-gave-us -some-good-genes-and-nasty-diseases-180960870.

8. Simon Szreter, "Chapter One: (The Wrong Kind of) Gonorrhea in Antiquity," in *The Hidden Affliction: Sexually Transmitted Infections and Infertility in History* (Woodbridge, UK: Boydell & Brewer Ltd., 2019), www.ncbi.nlm.nih .gov/books/NBK547155.

9. F. Gruber, J. Lipozenčić, and T. Kehler, "History of Venereal Diseases from Antiquity to the Renaissance," *Acta Dermatovenerologica Croatica: ADC* 23, no. 1 (2015): 1–11.

10. Kachiu C. Lee, "The Clap Heard Round the World," *Archives of Dermatology* 148, no. 2 (2012): 223, https://doi.org/10.1001/archdermatol.2011.2716.

11. Kate Lister, "The Bishop's Profitable Sex Workers," Wellcome Collection, June 5, 2018, https://wellcomecollection.org/articles /WxEniCQAACQAvmUE.

12. John Frith, "Syphilis—Its Early History and Treatment Until Penicillin and the Debate on Its Origins," *Journal of Military and Veteran Health* 20, no. 4, https://jmvh.org/article/syphilis-its-early-history-and-treatment-until -penicillin-and-the-debate-on-its-origins.

13. Sonny Maley, "Syphilis—What's in a Name?," University of Glasgow Library, September 12, 2014, https://universityofglasgowlibrary.wordpress .com/2014/09/12/syphilis-whats-in-a-name.

14. S. Szreter, "Treatment Rates for the Pox in Early Modern England: A Comparative Estimate of the Prevalence of Syphilis in the City of Chester and Its Rural Vicinity in the 1770s," *Continuity and Change* 32, no. 2 (2017): 183–223, https://doi.org/10.1017/S0268416017000212.

15. Manjunath M. Shenoy, Amina Asfiya, and Malcolm Pinto, "'A Night with Venus, a Lifetime with Mercury': Insight into the Annals of Syphilis," *Archives of Medicine and Health Sciences* 6, no. 2 (2018): 290, https://doi .org/10.4103/amhs.amhs_131_18.

16. John Frith, "Syphilis—Its Early History and Treatment Until Penicillin," *JMVH* 20, no. 4 (November 2012), https://jmvh.org/article/syphilis-its -early-history-and-treatment-until-penicillin-and-the-debate-on-its-origins.

17. Linda Geddes, "The Fever Paradox," *New Scientist (1971)* 246, no. 3277 (2020): 39–41, www.ncbi.nlm.nih.gov/pmc/articles/PMC7195085.

18. I. M. Daey Ouwens et al., "Malaria Fever Therapy for General Paralysis of the Insane: A Historical Cohort Study," *European Neurology* 78 (2017): 56–62, www.karger.com/Article/FullText/477900#.

19. Matthew Gambino, "Fevered Decisions: Race, Ethics, and Clinical Vulnerability in the Malarial Treatment of Neurosyphilis, 1922–1953," *Hastings Center Report* 45, no. 4 (2015): 39–50, https://doi.org/10.1002/hast.451.

20. Predesh Parasseril Jose, Vatsan Vivekanandan, and Kunjumani Sobhanakumari, "Gonorrhea: Historical Outlook," *Journal of Skin and Sexually Transmitted Diseases* 2, no. 2 (2020): 110–114, https://jsstd.org/gonorrhea-historical-outlook.

21. Thomas Benedek, "History of the Medical Treatment of Gonorrhea," Antimicrobe, www.antimicrobe.org/h04c.files/history/Gonorrhea.asp.

22. J. L. Milton, *On the Pathology and Treatment of Gonorrhoea and Spermatorrhoea* (New York: William Wood & Company, 1887).

23. Jay Gladstein, "Hunter's Chancre: Did the Surgeon Give Himself Syphilis?," *Clinical Infectious Diseases* 41, no. 1 (2005): 128, https://doi.org/10.1086/430834.

24. George Qvist, "John Hunter's Alleged Syphilis," *Annals of the Royal College of Surgeons of England* 59, no. 3 (May 1977): 205–209.

25. Shane Seger, "Malarial Fever as Neurosyphilis Treatment: A Historical Case Study in Medical Ethics," Yale School of Medicine, April 7, 2015, https://medicine.yale.edu/news-article/malarial-fever-as-neurosyphilis-treatment-a-historical-case-study-in-medical-ethics.

26. Jason Daley, "Why the Skeleton of the 'Irish Giant' Could Be Buried at Sea," *Smithsonian Magazine*, June 26, 2018, www.smithsonianmag.com/smart-news/why-skeleton-irish-giant-could-be-buried-sea-180969443.

27. "5 Things You Didn't Know About Burke & Hare," University of Edinburgh, January 14, 2021, www.ed.ac.uk/medicine-vet-medicine/postgraduate/postgraduate-blog/things-you-didnt-know-burke-hare.

28. A. W. Bates, "Dr Kahn's Museum: Obscene Anatomy in Victorian London," *Journal of the Royal Society of Medicine* 99, no. 12 (2006), www.ncbi.nlm.nih.gov/pmc/articles/PMC1676337.

29. Alex Schwartz, "What It Means to Have 'Undetectable' HIV—and Why You Need to Know," *Popular Science*, December 2, 2019, www.popsci.com/story/health/hiv-drugs-undetectable-status.

30. Laura Accinelli, "Where'd All Those Darn Hot Tubs Go?," *Los Angeles Times*, August 4, 1996, www.latimes.com/archives/la-xpm-1996-08-04-tm-31049-story.html.

31. Alex Schwartz, "How a Victorian Heart Medicine Became a Gay Sex Drug," *Popular Science*, June 28, 2019, www.popsci.com/wake-up-smell-the-poppers.

32. Chad C. Smith and Ulrich G. Mueller, "Sexual Transmission of Beneficial Microbes," *Trends in Ecology & Evolution* 30, no. 8 (June 27, 2015): 438–440, https://doi.org/10.1016/j.tree.2015.05.006.

33. Nancy A. Moran and Helen E. Dunbar, "Sexual Acquisition of Beneficial Symbionts in Aphids," *Proceedings of the National Academy of Sciences of the United States of America* 103, no. 34 (2006): 12803–12806, www.pnas.org /content/103/34/12803.

34. Claudia Damiani et al., "Paternal Transmission of Symbiotic Bacteria in Malaria Vectors," *Current Biology* 18, no. 23 (2008): R1087–R1088, https:// doi.org/10.1016/j.cub.2008.10.040.

35. Nirjal Bhattarai and Jack T Stapleton, "GB Virus C: The Good Boy Virus?," *Trends in Microbiology* 20, no. 3 (2012): 124–130, www.ncbi.nlm.nih .gov/pmc/articles/PMC3477489.

36. Ernest T. Chivero et al., "Human Pegivirus (HPgV; Formerly Known as GBV-C) Inhibits IL-12 Dependent Natural Killer Cell Function," *Virology* 485 (2015): 116–127, https://doi.org/10.1016/j.virol.2015.07.008.

7. HOW IS BABY FORMED?

1. Selena Simmons-Duffin, "The Texas Abortion Ban Hinges on 'Fetal Heartbeat.' Doctors Call That Misleading," NPR, September 3, 2021, www .npr.org/sections/health-shots/2021/09/02/1033727679/fetal-heartbeat-isnt -a-medical-term-but-its-still-used-in-laws-on-abortion.

2. "Ovism," Embryo Project Encyclopedia, accessed October 11, 2021, https://embryo.asu.edu/pages/ovism.

3. "Spermism," Embryo Project Encyclopedia, accessed October 11, 2021, https://embryo.asu.edu/pages/spermism.

4. "Aretaiou Kappadokou Ta sozomena = The Extant Works of Aretaeus, the Cappadocian," Internet Archive, https://archive.org/details /aretaioukappadok00aret.

5. "Works by Plato Circa 360 BC," JSTOR, www.jstor.org/stable/10.1525 /CA.2011.30.1.1.

6. Jen-Der Lee, "Childbirth in Early Imperial China," Institute of History and Philology, www2.ihp.sinica.edu.tw/file/2202dixRkWT.pdf.

7. "First English Book on Hysteria, 1603," British Library, www.bl.uk /collection-items/first-english-book-on-hysteria-1603.

8. "Conception and Childbirth," Encyclopedia.com, www.encyclopedia .com/history/news-wires-white-papers-and-books/conception-and-childbirth.

9. "Manuscript for the Health of Mother and Child," UCL, www.ucl.ac.uk /museums-static/digitalegypt/med/birthpapyrus.html.

10. Erin Beresini, "The Myth of the Falling Uterus," *Outside*, March 25, 2013, www.outsideonline.com/health/wellness/myth-falling-uterus.

11. Joseph Stromberg, "'Bicycle Face': A 19th-Century Health Problem Made Up to Scare Women Away from Biking," Vox, March 24, 2015, www .vox.com/2014/7/8/5880931/the-19th-century-health-scare-that-told-women -to-worry-about-bicycle.

12. "Uterine Prolapse," Cleveland Clinic, https://my.clevelandclinic.org /health/diseases/16030-uterine-prolapse.

13. "The Life and Works of Hildegard von Bingen (1098–1179)," Kenyon, www2.kenyon.edu/projects/margin/hildegar.htm.

14. Jen Gunter, "7 Fertility Myths That Belong in the Past," *New York Times*, April 15, 2020, www.nytimes.com/2020/04/15/parenting/fertility /trying-to-conceive-myths.html.

15. Jenny Morder, "What Science Says About Arousal During Rape," *Popular Science*, May 31, 2013, www.popsci.com/science/article/2013 -05/science-arousal-during-rape.

16. "The Curious Case of Mary Toft," University of Glasgow, www.gla .ac.uk/myglasgow/library/files/special/exhibns/month/aug2009.html.

17. Sabrina Imbler, "Why Historians Are Reexamining the Case of the Woman Who Gave Birth to Rabbits," Atlas Obscura, July 3, 2019, www .atlasobscura.com/articles/mary-toft-gave-birth-to-rabbits.

18. "1.5 The Theory of Maternal Impression," OpenLearn, www.open .edu/openlearn/ocw/mod/oucontent/view.php?id=65962&section=1.5.

19. F. Serpeloni et al., "Grandmaternal Stress During Pregnancy and DNA Methylation of the Third Generation: An Epigenome-Wide Association Study," *Translational Psychiatry* 7 (2017): e1202, https://doi.org/10.1038 /tp.2017.153.

20. C. Trompoukis et al., "Semen and the Diagnosis of Infertility in Aristotle," *Andrologia* 39, no. 1 (February 2007): 33–37, https://doi .org/10.1111/j.1439-0272.2006.00757.x.

21. Rebecca Flemming, "The Invention of Infertility in the Classical Greek World: Medicine, Divinity, and Gender," *Bulletin of the History of Medicine* 87, no. 4 (2013): 565–590, www.ncbi.nlm.nih.gov/pmc/articles/PMC 3904772.

22. J. K. Amory, "George Washington's Infertility: Why Was the Father of Our Country Never a Father?," *Fertility and Sterility* 81, no. 3 (2004): 495–499, https://pubmed.ncbi.nlm.nih.gov/15037389.

23. Fernando Tadeu Andrade-Rocha, "On the Origins of the Semen Analysis: A Close Relationship with the History of the Reproductive Medicine,"

Journal of Human Reproductive Sciences 10, no. 4 (2017): 242–255, www.ncbi .nlm.nih.gov/pmc/articles/PMC5799927.

24. J. Barkay and H. Zuckerman, "The Role of Cryobanking in Artificial Insemination," in *Human Artificial Insemination and Semen Preservation*, ed. Georges David and Wendel S. Price (New York: Plenum Press, 1980), https:// link.springer.com/chapter/10.1007/978-1-4684-8824-1_26.

8. HAVE WE ALWAYS USED BIRTH CONTROL?

1. Craig A. Hill, "The Distinctiveness of Sexual Motives in Relation to Sexual Desire and Desirable Partner Attributes," *Journal of Sex Research* 34, no. 2 (1997): 139–153, www.jstor.org/stable/3813561.

2. Janaka Bowman Lewis, *Freedom Narratives of African American Women: A Study of 19th Century Writings* (Jefferson, NC: McFarland & Company, 2017).

3. Fahd Khan et al., "The Story of the Condom," *Indian Journal of Urology* 29, no. 1 (2013): 12, https://doi.org/10.4103/0970-1591.109976.

4. "Charles Nelson Goodyear (1800–1860)," Museum of Contraception and Abortion, accessed October 11, 2021, https://muvs.org/en/topics/pioneers /charles-goodyear-1800-1860-en.

5. "Anthony Comstock's 'Chastity' Laws," PBS, accessed October 11, 2021, www.pbs.org/wgbh/americanexperience/features/pill-anthony -comstocks-chastity-laws.

6. Julie Morse, "Why Douching Won't Die," *Atlantic*, April 21, 2015, www .theatlantic.com/health/archive/2015/04/why-douching-wont-die/390198.

7. Amanda Jenkins, Deborah Money, and Kieran C. O'Doherty, "Is the Vaginal Cleansing Product Industry Causing Harm to Women?," *Expert Review of Anti-infective Therapy* 19, no. 3 (2020): 267–269, https://doi.org/10.108 0/14787210.2020.1822166.

8. "Can Coca-Cola Prevent Pregnancy?," Snopes.com, accessed October 11, 2021, www.snopes.com/fact-check/coca-cola-spermicide.

9. Ken Parejko, "Pliny the Elder's Silphium: First Recorded Species Extinction," *Conservation Biology* 17, no. 3 (2003): 925–927, https://doi .org/10.1046/j.1523-1739.2003.02067.x.

10. Zaria Gorvett, "The Mystery of the Lost Roman Herb," BBC, September 7, 2017, www.bbc.com/future/article/20170907 -the-mystery-of-the-lost-roman-herb.

11. John M. Riddle, *Contraception and Abortion from the Ancient World to the Renaissance* (Cambridge, MA: Harvard University Press, 1994), www.hup .harvard.edu/catalog.php?isbn=9780674168763; Helen King, "Eve's Herbs: A

History of Contraception and Abortion in the West," *Medical History* 42, no. 3 (1998): 412–414, www.ncbi.nlm.nih.gov/pmc/articles/PMC1044062.

12. E. R. Plunkett and R. L. Noble, "The Effects of Injection of Lithospermum Ruderale on the Endocrine Organs of the Rat," *Endocrinology* 49, no. 1 (1951): 1–7, https://doi.org/10.1210/endo-49-1-1.

13. "Medicinal Uses of Plants by Indian Tribes of Nevada," US National Library of Medicine, http://resource.nlm.nih.gov/10730550R; Ron D. Stubbs, "An Investigation of the Edible and Medicinal Plants Used by the Flathead Indians" (1966), *Graduate Student Theses, Dissertations, and Professional Papers* 6674, University of Montana ScholarWorks, https://scholarworks.umt.edu /cgi/viewcontent.cgi?article=7709&context=etd.

14. G. C. Jansen and Hans Wohlmuth, "Carrot Seed for Contraception: A Review," *Australian Journal of Herbal Medicine* 26 (2014): 10–17, www.research gate.net/publication/289343049_Carrot_seed_for_contraception_A_review.

15. R. Maurya et al., "Traditional Remedies for Fertility Regulation," *Current Medicinal Chemistry* 11, no. 11 (2004): 1431–1450, https://pubmed.ncbi .nlm.nih.gov/15180576.

16. Philippa Roxby, "Plant Chemicals Hope for 'Alternative Contraceptives,'" BBC, May 16, 2017, www.bbc.com/news/health-39923293.

17. Nadja Mannowetz et al., "Steroids and Triterpenoids Affect Sperm Fertility," *Proceedings of the National Academy of Sciences of the United States of America* 114, no. 22 (May 2017): 5743–5748, www.pnas.org/content/114/22/5743 .full.

18. "Wishbone Stem Pessary (Intracervical Device), Europe, 1880–1940," Science Museum Group, https://collection.sciencemuseumgroup .org.uk/objects/co96426/wishbone-stem-pessary-intracervical-device -europe-1880-1940-intra-uterine-device.

19. "Stem Pessary, Germany, 1925–1935," Wellcome Collection, https:// wellcomecollection.org/works/xkv7b3cb.

20. "Contraceptive Use in the United States by Method," Guttmacher Institute, May 2021, www.guttmacher.org/fact-sheet/contraceptive -method-use-united-states.

21. "The Dalkon Shield," Embryo Project Encyclopedia, https://embryo.asu .edu/pages/dalkon-shield.

22. David Hubacher, "The Checkered History and Bright Future of Intrauterine Contraception in the United States," *Perspectives on Sexual and Reproductive Health* 34, no. 2 (March/April 2002), www.gutt macher.org/journals/psrh/2002/03/checkered-history-and-bright-future -intrauterine-contraception-united-states.

23. Kai Bühling et al., "Worldwide Use of Intrauterine Contraception: A Review," *Contraception* 89 (2013), www.researchgate.net/publication/259086240_Worldwide_use_of_Intrauterine_Contraception_a_review.

24. "Mirena Lawsuits," Drugwatch, www.drugwatch.com/mirena/lawsuits.

25. Megan K. Donovan, "The Looming Threat to Sex Education: A Resurgence of Federal Funding for Abstinence-Only Programs?," Guttmacher Institute, August 30, 2018, www.guttmacher.org/gpr/2017/03/looming-threat-sex-education-resurgence-federal-funding-abstinence-only-programs.

26. Anthony Paik, Kenneth J. Sanchagrin, and Karen Heimer, "Broken Promises: Abstinence Pledging and Sexual and Reproductive Health," *Journal of Marriage and Family* 78, no. 2 (2016): 546–561, https://doi.org/10.1111/jomf.12279.

27. Nikita Stewart, "Planned Parenthood in N.Y. Disavows Margaret Sanger over Eugenics," *New York Times*, July 21, 2020, www.nytimes.com/2020/07/21/nyregion/planned-parenthood-margaret-sanger-eugenics.html.

28. William Brangham, "How a Coney Island Sideshow Advanced Medicine for Premature Babies," PBS, July 21, 2015, www.pbs.org/newshour/health/coney-island-sideshow-advanced-medicine-premature-babies.

29. Amita Kelly, "Fact Check: Was Planned Parenthood Started to 'Control' the Black Population?," NPR, August 14, 2015, www.npr.org/sections/itsallpolitics/2015/08/14/432080520/fact-check-was-planned-parenthood-started-to-control-the-black-population.

30. Andrea DenHoed et al., "The Forgotten Lessons of the American Eugenics Movement," *New Yorker*, April 27, 2016, www.newyorker.com/books/page-turner/the-forgotten-lessons-of-the-american-eugenics-movement.

31. Pamela Verma Liao and Janet Dollin, "Half a Century of the Oral Contraceptive Pill: Historical Review and View to the Future," *Canadian Family Physician* 58, no. 12 (2012): e757–e760.

32. G. van N. Viljoen, "Plato and Aristotle on the Exposure of Infants at Athens," *Acta Classica* 2 (1959): 58–69, www.jstor.org/stable/24591098.

33. "Fact Check: Most Americans Do Not Oppose Abortion," *Reuters*, November 2, 2020, www.reuters.com/article/uk-factcheck-most-americans-not-anti-abo/fact-check-most-americans-do-not-oppose-abortion-idUSKBN27I2C0; "Food Security Status of U.S. Households with Children in 2020," Economic Research Service, US Department of Agriculture, www.ers.usda

.gov/topics/food-nutrition-assistance/food-security-in-the-us/key-statistics
-graphics.aspx#children.

34. Joan Alker and Alexandra Corcoran, "Children's Uninsured Rate Rises by Largest Annual Jump in More Than a Decade," Center for Children and Families, Georgetown University Health Policy Institute, October 8, 2020, https://ccf.georgetown.edu/2020/10/08/childrens-uninsured-rate-rises-by -largest-annual-jump-in-more-than-a-decade-2.

35. P. Wilson-Kastner and B. Blair, "Biblical Views on Abortion: An Episcopal Perspective," *Conscience* 6, no. 6 (1985): 4–8, https://pubmed.ncbi.nlm .nih.gov/12178933.

36. Rachel Mikva, "There Is More Than One Religious View on Abortion—Here's What Jewish Texts Say," Conversation, updated September 7, 2021, https://theconversation.com/there-is-more-than-one-religious-view -on-abortion-heres-what-jewish-texts-say-116941.

37. Malcolm Potts, Maura Graff, and Judy Taing, "Thousand-Year-Old Depictions of Massage Abortion," Bixby Center, 2007, http://bixby.berkeley .edu/wp-content/uploads/2015/03/Thousand-year-old-depictions-of-massage -abortion.pdf.

38. Andrea Zlotucha Kozub, "'To Married Ladies It Is Peculiarly Suited': Nineteenth-Century Abortion in an Archaeological Context," *Historical Archaeology* 52, no. 2 (April 2018), www.researchgate.net /publication/324514590_To_Married_Ladies_It_Is_Peculiarly_Suited _Nineteenth-Century_Abortion_in_an_Archaeological_Context.

39. "Selling Sex," *Capitalism by Gaslight: The Shadow Economies of 19th-Century America*, https://librarycompany.org/shadoweconomy/section4_13 .htm.

40. Ryan Johnson, "A Movement for Change: Horatio Robinson Storer and Physicians' Crusade Against Abortion," *James Madison Undergraduate Research Journal* 4, no. 1 (2017): 13–23, http://commons.lib.jmu.edu/jmurj/vol4/iss1/2.

41. Kristina Killgrove, "Aborted Fetus and Pill Bottle in 19th Century New York Outhouse Reveal History of Family Planning," *Forbes*, April 28, 2018, www.forbes.com/sites/kristinakillgrove/2018/04/20/aborted-fetus-and-pill -bottle-in-19th-century-new-york-outhouse-reveal-history-of-family -planning/?sh=6315ba5575a1.

42. "A Study of Abortion in Primitive Societies: A Typological, Distributional, and Dynamic Analysis of the Prevention of Birth in 400 Preindustrial Societies by Devereux, George, 1908–1985," Internet Archive, https://archive .org/details/studyofabortioni00deve/page/n15/mode/2up.

9. WHY DON'T OUR BODIES ALWAYS COOPERATE WITH OUR HORNY HEARTS?

1. Caroline Moreau, Anna E Kågesten, and Robert Wm Blum, "Sexual Dysfunction Among Youth: An Overlooked Sexual Health Concern," *BMC Public Health* 16, no. 1 (2016), https://doi.org/10.1186/s12889-016-3835-x.

2. Mats Holmberg, Stefan Arver, and Cecilia Dhejne, "Supporting Sexuality and Improving Sexual Function in Transgender Persons," *Nature Reviews Urology* 16, no. 2 (2018): 121–139, https://doi.org/10.1038/s41585-018-0108-8.

3. J. Shah, "Erectile Dysfunction Through the Ages," *BJU International* 90 (2002): 433–441, https://doi.org/10.1046/j.1464-410X.2002.02911.x.

4. A. A. Shokeir and M. I. Hussein, "Sexual Life in Pharaonic Egypt: Towards a Urological View," *International Journal of Impotence Research* 16, no. 5 (2004): 385–388, https://doi.org/10.1038/sj.ijir.3901195.

5. Candida Moss, "From Foods That Make You Fart to Bull Urine Ointment, How the Ancients Dealt with Man's Struggle to Get It Up," Daily Beast, July 14, 2019, www.thedailybeast.com/impotency-how-the-ancient -world-dealt-with-mans-struggle-to-get-it-up.

6. Aslam Farouk-Alli and Mohamed Shaid Mathee, "The Tombouctou Manuscript Project: Social History Approaches," in *The Meanings of Timbuktu*, ed. Shamil Jeppie and Souleymane Bachir Diagne (Cape Town: HSRC Press in association with CODESRIA, 2008), https://codesria .org/IMG/pdf/The_Meanings_of_Timbuktu_-_Chapter_12_-_The _Tombouctou_Manuscript_Project__social_history_approaches.pdf.

7. Jacqueline Murray, "On the Origins and Role of 'Wise Women' in Causes for Annulment on the Grounds of Male Impotence," *Journal of Medieval History* 16, no 3 (1990): 235–249, www.sciencedirect.com/science/article /abs/pii/030441819090004K.

8. "Between a Rock and a Hard Place: Impotence Tests in the Middle Ages," *Whores of Yore*, July 25, 2017, www.thewhoresofyore.com/katersquos-journal /between-a-rock-and-a-hard-place-impotence-tests-in-the-middle-ages.

9. "St. Albertus Magnus," *Encyclopaedia Britannica*, accessed October 11, 2021, www.britannica.com/biography/Saint-Albertus-Magnus.

10. "John R. Brinkley," Kansas Historical Society, April 2014, www.kshs .org/kansapedia/john-r-brinkley/11988.

11. Matthew Wills, "This Doc Was Really Nuts," JSTOR Daily, July 28, 2016, https://daily.jstor.org/this-doc-was-really-nuts.

12. Eric Grundhauser, "The True Story of Dr. Voronoff's Plan to Use Monkey Testicles to Make Us Immortal," Atlas Obscura, February 29, 2016, www

.atlasobscura.com/articles/the-true-story-of-dr-voronoffs-plan-to-use
-monkey-testicles-to-make-us-immortal.

13. M. A. Buchholz and M. Cervera, "Radium Historical Items Catalog: Final Report," US Nuclear Regulatory Commission, August 2008, www.nrc.gov/docs/ML1008/ML100840118.pdf.

14. Jacob Kan et al., "Biographical Sketch: Giles Brindley, FRS," *British Journal of Neurosurgery* 28, no. 6 (2017): 704–706, www.tandfonline.com/doi/full/10.3109/02688697.2014.925085.

15. Laurence Klotz, "How (Not) to Communicate New Scientific Information: A Memoir of the Famous Brindley Lecture," *BJU International*, October 13, 2005, https://bjui-journals.onlinelibrary.wiley.com/doi/abs/10.1111/j.1464-410X.2005.05797.x.

16. G. S. Brindley, "The Fertility of Men with Spinal Injuries," *Spinal Cord* 22 (1984): 337–349, www.nature.com/articles/sc198456.

17. Ian Osterloh, "How I Discovered Viagra: Drug for Heart Disease Revealed an Unrelated Side Effect," *Cosmos*, April 27, 2015, https://cosmosmagazine.com/biology/how-i-discovered-viagra.

18. Alex Schwartz, "How a Victorian Heart Medicine Became a Gay Sex Drug," *Popular Science*, June 28, 2019, www.popsci.com/wake-up-smell-the-poppers.

19. Deborah Netburn, "Viagra for Women? Blue Pills May Help Alleviate Menstrual Cramps," *Los Angeles Times*, December 9, 2013, www.latimes.com/science/sciencenow/la-sci-sn-viagra-for-women-menstrual-cramps-20131209-story.html.

20. Christopher Ingraham, "The Military Spends Five Times As Much on Viagra as It Would on Transgender Troops' Medical Care," *Washington Post*, July 26, 2017, www.washingtonpost.com/news/wonk/wp/2017/07/26/the-military-spends-five-times-as-much-on-viagra-as-it-would-on-transgender-troops-medical-care.

21. "Addyi Is Not a 'Female Viagra,' but It Can Open an Important Discussion," Harvard Health Publishing, October 21, 2015, www.health.harvard.edu/womens-health/addyi-is-not-a-female-viagra-but-it-can-open-an-important-discussion.

22. Angela Chen, "'Female Viagra' Is Back and Easily Available Online—Which Means It Could Be More Harmful Than Ever," *The Verge*, June 13, 2018, www.theverge.com/2018/6/13/17458608/female-viagra-addyi-flibanserin-sex-fda-health.

23. Nicole Blackwood, "FDA Approves Vyleesi, a New 'Female Viagra.' What Issues Can It Actually Solve?," *Chicago Tribune*, June 24, 2019, www

.chicagotribune.com/lifestyles/health/ct-life-female-viagra-fda-tt-20190624
-20190624-76yhvyznpjbhtoydiygalx32g4-story.html.

10. WHAT IS PORN, EXACTLY?

1. "The Earliest Pornography?," *Science*, May 13, 2009, www.sciencemag
.org/news/2009/05/earliest-pornography.

2. A. Verit, "Recent Discovery of Phallic Depictions in Prehistoric Cave
Art in Asia Minor," *European Urology Supplements* 16, no. 3 (2017), https://doi
.org/10.1016/s1569-9056(17)30572-9.

3. Ogden Goelet, "Nudity in Ancient Egypt," *Source: Notes in the History of
Art* 12, no. 2 (1993): 20–31, https://doi.org/10.1086/sou.12.2.23202932.

4. Gay Robins, *Women in Ancient Egypt* (Cambridge, MA: Harvard University Press, 1993), https://archive.org/details/womeninancienteg00robi.

5. Ann Babe, "Object of Intrigue: Moche Sex Pots," Atlas Obscura, March
8, 2016, www.atlasobscura.com/articles/object-of-intrigue-moche-sex-pots.

6. Mary Weismantel, "Moche Sex Pots: Reproduction and Temporality in
Ancient South America," *American Anthropologist* 106, no. 3 (2004): 495–505,
www.faculty.fairfield.edu/dcrawford/weismantel.pdf.

7. Isaac Stone Fish, "Does Japan's Conservative Shinto Religion Support Gay Marriage?," *Foreign Policy*, June 29, 2015, https://foreignpolicy
.com/2015/06/29/what-does-japan-shinto-think-of-gay-marriage; Brian Ash-
craft, "Vagina Artist Arrested in Japan," *Kotaku*, July 14, 2014, https://kotaku
.com/vagina-artist-arrested-in-japan-1604550217.

8. David Mikkelson, "The Love Machine," Snopes.com, www.snopes.com
/fact-check/the-love-machine.

9. Ian Buruma, "The Joy of Art: Why Japan Embraced Sex with a Passion,"
Guardian, September 27, 2013, www.theguardian.com/artanddesign/2013
/sep/27/joy-art-japan-sex-passion.

10. Alastair Sooke, "Sexually Explicit Japanese Art Challenges
Western Ideas," BBC, October 10, 2014, www.bbc.com/culture
/article/20131003-filth-or-fine-art.

11. "Japan: Possession of Child Pornography Finally Punishable," Library of Congress, August 4, 2014, www.loc.gov/item/global-legal-monitor
/2014-08-04/japan-possession-of-child-pornography-finally-punishable.

12. Cameron W. Barr, "Why Japan Plays Host to World's Largest Child
Pornography Industry," *Christian Science Monitor*, April 2, 1997, www.cs
monitor.com/1997/0402/040297.intl.intl.1.html.

13. Annetta Black, "Gabinetto Segreto," Atlas Obscura, February 14, 2011,
www.atlasobscura.com/places/gabinetto-segreto.

14. John R. Clarke, "Before Pornography: Sexual Representation in Ancient Roman Visual Culture," in *Pornographic Art and the Aesthetics of Pornography*, ed. Hans Maes (London: Palgrave Macmillan, 2013), https://link.springer.com/chapter/10.1057%2F9781137367938_8.

15. "29 Things You (Probably) Didn't Know About the British Museum," British Museum, September 7, 2021, https://blog.britishmuseum.org/29-things-you-probably-didnt-know-about-the-british-museum.

16. Joseph Price et al., "How Much More XXX Is Generation X Consuming? Evidence of Changing Attitudes and Behaviors Related to Pornography Since 1973," *Journal of Sex Research* 53, no. 1 (2015): 12–20, https://doi.org/10.1080/00224499.2014.1003773.

17. Ralph Blumenthal, "'Hard-Core' Grows Fashionable—and Very Profitable," *New York Times*, January 21, 1973, www.nytimes.com/1973/01/21/archives/pornochic-hardcore-grows-fashionableand-very-profitable.html.

11. WHY DO SO MANY OF US LIKE SEX THAT ISN'T "NORMAL"?

1. Christian Joyal and Julie Carpentier, "The Prevalence of Paraphilic Interests and Behaviors in the General Population: A Provincial Survey," *Journal of Sex Research* 54 (2017): 161–171, www.researchgate.net/publication/289368254_The_Prevalence_of_Paraphilic_Interests_and_Behaviors_in_the_General_Population_A_Provincial_Survey.

2. Adee Braun, "The Once-Common Practice of Communal Sleeping," Atlas Obscura, June 26, 2017, www.atlasobscura.com/articles/communal-sleeping-history-sharing-bed.

3. Harry Oosterhuis, *Stepchildren of Nature: Krafft-Ebing, Psychiatry, and the Making of Sexual Identity* (Chicago: University of Chicago Press, 2000).

4. Sarah J. Jones, Caoilte Ó Ciardha, and Ian A. Elliott, "Identifying the Coping Strategies of Nonoffending Pedophilic and Hebephilic Individuals from Their Online Forum Posts," *Sexual Abuse* 33, no. 7 (2021): 793–815, https://journals.sagepub.com/doi/full/10.1177/1079063220965953; Luke Malone, "You're 16. You're a Pedophile. You Don't Want to Hurt Anyone. What Do You Do Now?," Medium, August 11, 2014, https://medium.com/matter/youre-16-youre-a-pedophile-you-dont-want-to-hurt-anyone-what-do-you-do-now-e11ce4b88bdb.

5. Shayla Love, "Pedophilia Is a Mental Health Issue. It's Still Not Treated as One," *Vice*, August 24, 2020, www.vice.com/en/article/y3zk55/pedophilia-is-a-mental-health-issue-its-still-not-treated-as-one.

6. Neuroskeptic, "The Erogenous Zones of the Brain," *Discover*, September 7, 2013, www.discovermagazine.com/mind/the-erogenous-zones-of-the-brain.

7. Daniil Ryabko and Zhanna Reznikova, "On the Evolutionary Origins of Differences in Sexual Preferences," *Frontiers in Psychology* 6 (2015), https://doi.org/10.3389/fpsyg.2015.00628.

8. J. Bivona and J. Critelli, "The Nature of Women's Rape Fantasies: An Analysis of Prevalence, Frequency, and Contents," *Journal of Sex Research* 46, no. 1 (2009): 33–45, https://pubmed.ncbi.nlm.nih.gov/19085605.

9. Mark Hay, "Fantasies of Forced Sex Are Common. Do They Enable Rape Culture?," Aeon, June 3, 2019, https://aeon.co/ideas/fantasies-of-forced-sex-are-common-do-they-enable-rape-culture.

10. Matt Lebovic, "When Israel Banned Nazi-Inspired 'Stalag' Porn," *Times of Israel*, November 17, 2016, www.timesofisrael.com/when-israel-banned-nazi-inspired-stalag-porn.

11. Andrew O'Hehir, "Israel's Nazi-Porn Problem," Salon, April 11, 2008, www.salon.com/2008/04/11/stalags.

12. "National Library of Israel's Hidden 'Stalag' Collection," Atlas Obscura, www.atlasobscura.com/places/national-library-of-israel-s-hidden-stalag-collection.

13. Samir S. Patel, "How the Leatherdykes Helped Change Feminism," Atlas Obscura, May 1, 2017, www.atlasobscura.com/articles/leather-feminism-lesbian-leatherdyke-bdsm.

14. Nadja Spiegelman, "James Joyce's Love Letters to His 'Dirty Little Fuckbird,'" *Paris Review*, February 2, 2018, www.theparisreview.org/blog/2018/02/02/james-joyces-love-letters-dirty-little-fuckbird.

15. M. D. Griffiths, "Eproctophilia in a Young Adult Male," *Archives of Sexual Behavior* 42 (2013): 1383–1386, https://doi.org/10.1007/s10508-013-0156-3.

16. "Ancient History Sourcebook: Suetonius: De Vita Caesarum—Nero, c. 110 C.E.," Fordham University, https://sourcebooks.fordham.edu/ancient/suet-nero-rolfe.asp.

17. Leo Damrosch, "Friends of Rousseau," *Humanities* 33, no. 4 (July/August 2012), www.neh.gov/humanities/2012/julyaugust/feature/friends-rousseau.

Rachel Feltman's first paying gig was organizing a bookshelf full of textbooks on vulvar disease at the age of seven, and she never looked back. She's executive editor of *Popular Science* and hosts *PopSci*'s podcast *The Weirdest Thing I Learned This Week*. In 2014, Feltman founded the *Washington Post*'s *Speaking of Science* blog, known for headlines like "You Probably Have Herpes, but That's Really Okay" and "Uranus Might Be Full of Surprises." Feltman studied environmental science at Simon's Rock and has a master's in science reporting from New York University. She's a musician, an actress, and the stepmom of a very spry fifteen-year-old cat.